THE REAL ADAM

Reconciling Evolution with the Fall of Man and Original Sin

David Korotney

THE REAL ADAM

Cover design by Kevin Todd O'Ray.

Table of Contents

Introduction

In *The Problem of Pain*, C.S. Lewis addresses the tendency of some Christians to treat the Bible as a complete compendium of knowledge, able to answer all questions about the natural world. He retorts by saying:

> *But it is nothing of the sort; the curtain has been rent at one point, and at one point only, to reveal our immediate practical necessities and not to satisfy our intellectual curiosities.*

The immediate practical necessities of which Lewis speaks surround man's deficient moral state before the holy God and his need for a Savior. Everything else is an intellectual curiosity. Galileo, the 16th/17th Century astronomer and inventor, stated it more poetically:

> *The Bible shows the way to go to heaven, not the way the heavens go. (letter to Madame Christina of Lorraine, Grand Duchess of Tuscany)*

Biological evolution and its compatriot deep time are arguably intellectual curiosities, things we need not know anything about in order to have a relationship with God. As such, we should not expect the Bible to have anything to say about them. What the Bible does say about the origin of the universe and humanity is provided not in the language of physics and chemistry, but rather as a saga of God's intentions, our failings, and the path to redemption. It is a story of people, not things.

Even so, if one chooses to accept evolution as a scientific fact, it has some significant implications for Christian doctrine. Aside from the conspicuous fact that evolution's millions and billions of years contrasts rather dramatically with the Bible's story of a six-day creation, evolution also negates any possibility that life on earth was originally perfect, free from all suffering and death. A more serious implication of evolution is that Adam and Eve were not the first human beings to have existed nor the only human beings on earth at the time of the events in the garden of Eden. This state of affairs disrupts the traditional belief that the sin of Adam and Eve created within them a new, irresistible inclination towards sin that was passed on to all their progeny, and thus to every single human being who has ever existed.

Many others have taken on the task of defending the science behind evolution and identifying areas of compatibility between it and the Biblical account of creation in Genesis Chapters 1 and 2. Less thoroughly addressed is the issue of how humans, being the products of evolution, can be sinners who are in need of a Savior. This is the alpha question, the one on which the success or failure of any attempt to reconcile evolution with orthodox Christian doctrine turns. There are also several related questions that must be answered, such as how humans could be created in God's image while nevertheless being imperfect, and how a good God can create a world where suffering and death are commonplace.

What follows is an alternative paradigm for understanding the origin of not only the universe, life, and mankind, but most particularly the origin of sin. This "Revised Origins Paradigm" asserts that all life on earth arose from an evolutionary process spanning billions of years, that suffering and death were an inherent aspect of that process, and that Adam and Eve were ordinary imperfect humans with a special calling to serve as God's emissaries to the rest of humanity. It is contrasted with the "Traditional Origins Paradigm" which takes a more literal view of the Genesis creation account and holds that God initially created the world in a state of perfection that was free from all suffering and death, and that Adam and Eve were the first two human beings to have been created and were perfect physically, intellectually, and morally. Both the Traditional and Revised Origins Paradigms treat the disobedience of Adam and Eve as having had a profound and negative impact on all humanity, but precisely what this means differs between the two paradigms. Specifically, the concept of original sin as a corruption of a formally perfect human nature which was then passed on to all of Adam and Eve's progeny is something that necessarily needs modification if there is to be any hope of harmonizing evolution with Christianity.

This book is divided into four parts. Part I describes the state of the world before the appearance of Adam and Eve, demonstrating why an imperfect world resulting from an evolutionary process is consistent with the Biblical description and God's grand purposes. Part II then walks through the story of Adam and Eve, from their "creation" through the standard procreative route somewhere in ancient Mesopotamia, to their arrival in the garden of Eden and God's purposes for leading them there, and finally the nature of their disobedience to God's command and God's punishments. Part III focuses on the doctrine of original sin, beginning with the many and disparate views of the moral state of man before and after the Fall of man, Augustine's formulation and its weaknesses, the complex nature of sin, and most importantly how it can be that mankind can be sinners in need of a Savior if they did not fall into depravity from some previously perfect moral state. Finally, Part IV puts the pieces together to summarize the Revised Origins Paradigm and describe how it is not only consistent with orthodox Christian dogma concerning sin and salvation, but does a better job than the traditional approach at accounting for the empirical evidence concerning the disordered state of the universe and human behavior.

Because the Traditional Origins Paradigm places an emphasis on ideas that have been held since antiquity, this book likewise makes prevalent use of the views of theologians from the first five centuries of the current era, that period of Christendom when the core doctrines were developed. This includes, among others, Justin Martyr, Theophilus of Antioch, Irenaeus of Lyons, Clement of Alexandria, Tertullian, Athanasius of Alexandria, the Cappadocian Fathers, Ambrose of Milan, Cyril of Alexandria, and of course Augustine of Hippo who is generally regarded as the architect of the doctrine of original sin that is now accepted among western churches. Many elements of the Revised Origins Paradigm have their root in ideas expressed by these early theologians.

The Revised Origins Paradigm also owes much to those modern scholars who have studied the issues with Christian doctrine that arise if evolution is taken seriously. Those whose work most directly contributed to this book are listed in the References section. In many ways, the Revised Origins Paradigm is not new, but rather provides a single framework within which their ideas find interconnection.

In general, quoted Bible passages are taken from the New International Version. Occasionally, other translations are used when the NIV may not capture some nuance in the original Hebrew or Greek that may be important in the rationale for the Revised Origins Paradigm. Also, portions of Bible verses are sometimes quoted rather than the full verse in order to omit elements that may not be directly relevant to the point at hand. In such cases the Bible reference is amended by "a" to indicate that the quoted portion is at the beginning of the full verse or "b" to indicate that the quoted portion is at the end of the full verse.

Part I

Creation Before the Fall

Chapter 1
Biblical Literalism and the Creation Account

This is not a book about the Biblical creation account in Genesis Chapters 1 and 2. How the text could be read in a less literal way to account for deep time (millions and billions of years) and biological evolution has been investigated by many others who have delved into linguistic style, genre, diction, and cultural context. Instead, this book is focused on how the story of Adam and Eve in Genesis Chapters 2 and 3 can be true if they emerged from an evolutionary process, and most critically how it is that the Fall could in that case have resulted in the universality of sin that exists today and man's unqualified need for a Savior.

Nevertheless, the picture would not be complete without some discussion of the metaphorical and imprecise language used throughout the Bible, and how in light of that nonliteral language it is legitimate to also approach the Genesis creation account in a nonliteral way. Moreover, there are significant practical problems created by a purely literal reading of the Genesis creation account, leading even literalists to formulate supplemental narratives that are in some cases highly improbable. These factors pave the way for the discussion in subsequent chapters about how Biblical account of Adam and Eve can be true even if they were the products of biological evolution.

The trouble with strict literalism

Literalism is an approach to Biblical interpretation which treats every Bible word, phrase, and description as being true in its most literal, face-value, or primary sense unless there are unambiguous indications within the text that it should be treated otherwise.

Literalists will sometimes refer to this approach as "straight reading" or "plain reading." A plain reading of the Bible is one that presumes no underlying meaning which takes precedence, no allegory of spiritual truth told in natural terms, no poetic or artistic renderings of facts, no pseudo-history, and no creative embellishments. The text is taken unquestioningly at face value, its first and most easily derived meaning assumed to be the only meaning that truly matters (though some literalists will accede to additional, nonliteral meanings so long as the literal meaning is retained as preeminent). Extra-Biblical considerations are generally abhorred, or at least strongly disfavored.

Biblical literalists tend also to be strict inerrantists, believing that the Bible does not and cannot make any false or misleading statements about anything. On any topic on which the Bible touches - not only matters of faith, but also people, places, times, ages, quotations, events, explanations - it is 100% true down to the smallest detail. This belief derives from the view, commonly held among even Christians who are not inerrantists, that Scripture is the inspired Word of God. Inerrantists argue that God would not have permitted errors in so important a product as the Bible. The fact that God worked through sinful people who had their own personalities, writing styles, cultural influences, biases, and limited understanding is, in the view of inerrantists, irrelevant. In His mysterious way, God would have ensured that the words He wanted written would be written and would not have permitted the human instruments He used to distort the inspiration He provided, intentionally or unintentionally.

However, the more widely held view among Christians is that the Bible is infallible in matters of faith but is not necessarily true in every detail. This view is based on the expectation that it would only have been matters of faith that concerned God when He provided inspiration to the Bible's authors. God would not have stopped His human instruments from saying things in their own ways or from relating events based on their own faulty memories or faulty understanding of the world. To think otherwise would be to believe in a sort of Holy Spirit possession wherein the individual losses all autonomy, an idea at odds with belief in a God who honors free will at all times.

There is little evidence that inerrancy and its brother literalism were commonly held views of the 66 books of the Bible at the time each book was written. Instead, these views developed considerably later. As discussed further below, many theologians in the first few centuries CE took nonliteral views of many parts of Scripture, including the creation account in the first several chapters of Genesis.

More importantly, literalism is more problematic than its adherents generally admit. To begin with, literalists tend to be unyielding and elitist in their views, which does little to attract non-Christians to the faith. Typical of their hardline views are statements such as this from James Quiggle:

> *If one believes in inspiration, then one must believe that the narratives concerning Adam and Eve are historically factual.*

Henry M. Morris goes further, saying that those unwilling the accept six literal days of creation are in serious spiritual jeopardy:

> *If we are going to call Him a liar, we must be willing to face the consequences of blasphemy.*

Faced with a choice between being called a blasphemer for reading the Genesis creation account in a less than fully literal way, or rejecting faith in Jesus altogether because of the narrow-minded and unyielding positions of some Christians, some nonbelievers may unfortunately take the latter route.

To their credit, literalists take this approach out of both a respect for God's Word and a concern that anything less could lead to a dilution or even nullification of the core Christian message. Mortenson and Ury put it this way:

> *If we reinterpret the Scriptures on the creation days, Fall, and Flood, where will the process stop? Why stop with Genesis? What about the virgin birth or Resurrection of Jesus Christ?*

This is a legitimate, if exaggerated, concern. Theologians who question the literal historicity of the origins narrative in Genesis typically are not out to do the same with the Gospel stories. That others may make the extrapolation feared by Mortenson and Ury is not reason enough to adopt a purely literal approach to the Biblical origins narrative.

A weightier concern with literalism is that it creates a set of expectations that cannot be met in practice. In addition to the fact that literalists themselves do not apply their principles of interpretation consistently across all of Scripture - they actually do make nonliteral interpretations in many cases even when the text provides no obvious basis for it - they also maintain false notions of the immutability of doctrine and the subservience of all facts, scientific or otherwise, to their counterparts in Scripture.

Literalists often begin with an appeal to genre: the Genesis origins narrative should be read as literal history because it is written as if it were literal history, with linguistic markers and referents which are indicative of historical narratives. But not all scholars agree. Some compare it to creation narratives from other cultures and see little difference in style, thus declaring the Biblical version a mythic portrayal having little to do with what actually occurred. Others agree that it has enough linguistic indicators of historical narrative to validate treating it as such without it necessarily being literally true in every detail. Still others see it as describing only a spiritual history but in natural terms familiar to the original audience.

Literalists also point to the fact that other books in the Bible make reference to Adam and Eve, Cain and Abel, God's role in creation, the serpent, the institution of marriage, Adam's sin, Noah, and the great flood, by and large assuming that these people and events represent actual history. If the authors of these other books and the people described in them believed that the Genesis origins narrative was literal history, so the argument goes, then we must as well if we are to be faithful to our belief in the Bible as God's inspired word.

But what else should we expect Biblical personages to have believed? There was nothing like today's sciences to tell them otherwise and no such thing as linguistic analysis. The only alternative to the origins narrative in Genesis was the narratives from surrounding cultures which they would have rejected on the grounds that they supported a rival religion and gods who were not gods. They quite literally had no basis on which and no incentive to treat the Biblical origins narrative as anything other than literal history.

More importantly, we know for a fact that Biblical personages were wrong in their beliefs on numerous occasions. Abraham believed he could usher in God's promise by sleeping with his wife's handmaid. Moses believed he was an inadequate speaker. Ezekiel thought he was the only faithful Jew left on earth. Jonah believed he could outrun God. Joseph believed that his fiancé Mary had slept with another man.

But surely references to the origins narrative made by Jesus must be treated differently, literalists contend. Jesus would have known if they were in any way false or inaccurate - Jesus could not have *not* known. Except that it is indeed possible that Jesus did not know that the origins narrative was something less than literally true in every detail. We know this is a possibility because Jesus admitted that there are some things He did not know, as the following examples demonstrate:

> *At once Jesus realized that power had gone out from him. He turned around in the crowd and asked, "Who touched my clothes?" (Mark 5:30)*
>
> *But about that day or hour no one knows, not even the angels in heaven, nor the Son, but only the Father. (Mark 13:32)*

Jesus's knowledge had self-imposed limitations. While we do not know the precise extent of those limitations, we cannot rule out the possibility that Jesus maintained the common understanding of the Genesis creation account held by Jews of His time in order to best operate within that culture.

The figurative interpretations of literalists

Despite touting a literal approach to Biblical interpretation, literalists nevertheless take a figurative approach to many aspects of the Genesis creation account. For instance, they will often point to the fact that God lacks a physical body and therefore could not literally have breathed into Adam's nostrils, or used His hands to plant a garden or make garments from animal skins. They likewise say that God could not literally have rested on the seventh day since He needs no rest, and that the serpent could not have literally eaten dust since it contains no nutritional value.

One noteworthy deviation that literalists take from their own interpretive approach is in regards to the serpent which tempts Eve in Genesis Chapter 3. The serpent is described as being

> *more crafty than any of the wild animals the LORD God had made. (Genesis 3:1a)*

God's curse on the serpent not only reiterates its animal nature, but includes punishments that would be fitting for a physical animal.

> *Cursed are you above all livestock and all wild animals. You will crawl on your belly and you will eat dust all the days of your life. (Genesis 3:14b)*

The same Hebrew word translated as serpent in the Genesis account (*nachash*) is used elsewhere in the Bible, and nearly always represents a snakelike animal.

> *Then the LORD sent venomous snakes [nachash] among them; they bit the people and many Israelites died. (Numbers 21:6, brackets added)*

> *It will be as though a man fled from a lion only to meet a bear, as though he entered his house and rested his hand on the wall only to have a snake [nachash] bite him. (Amos 5:19, brackets added)*

The plain reading of the Genesis story is that the serpent was an animal like any other animal, albeit one with intelligence and the ability to speak. There is no explicit statement in Genesis or elsewhere in the Bible to say otherwise. Nevertheless, literalists agree with nearly all other theologians that Eve was interacting not with an animal per se but rather with the devil who was animating or appearing to animate the serpent's mouth. They conclude this not because there is any statement in the Bible to say so, but because there is no evidence than any animal has the intelligence and ability to speak (an extra-Biblical fact) and because the Genesis story could not otherwise be made consistent with the redemptive story of Christ.

Literalists also take a creative approach to maintaining a consistent and plain-reading treatment of "day" throughout the Genesis creation account. They begin by asserting that each of the days of creation in Genesis Chapter 1 indicate a single 24-hour period, such that all that exists was created out of nothing in six literal days. But this presents a problem when the exact same Hebrew word for "day" (*yom*) is used in God's warning to Adam.

> *but of the tree of the knowledge of good and evil you shall not eat, for in the day that you eat of it you shall surely die. (Genesis 2:17, English Standard Version)*

Adam did not physically die on the day that he ate from the forbidden tree, but lived at least 800 centuries more. Rather than conceding that "day" in God's warning - and thus also in the Genesis 1 creation account - should be interpreted more figuratively to indicate a time period much longer than 24 hours, they conclude that in some sense Adam began to physically die on the day that he ate from the tree of the knowledge of good and evil. Some go so far as to say that, consistent with 2 Peter 3:8, the "day" to which God referred in His warning really meant 1,000 years, or, consistent with 1 Kings 2:42, that "in the day" was an idiom merely indicating the certainty of death rather than its timing. None of these are consistent with a "plain reading" of the Bible.

Literalists also point to spiritual death rather than or in addition to physical death as a possible meaning in God's warning. But the most straightforward reading of God's warning, and thus the most literal reading, is that He meant physical death, based on God's later curse on Adam.

> *for dust you are and to dust you will return (Genesis 3:19b)*

As will be discussed in Chapter 5, there are good reasons to believe that God's warning to Adam did in fact mean something other than physical death within 24 hours of eating from the forbidden tree. Thus literalists have properly identified a case in which it is both necessary and proper to interpret a passage in a less than purely literal fashion despite the fact that there is no unambiguous indication in the text itself that it was intended to be figurative.

The limits of sola scriptura

The interpretation of one part of the Bible based on its connections to other parts of Scripture is a solid and a commonly used method of interpretation. However, it is taken to the extreme with literalists, who generally exclude all information from outside of the Bible based on the presupposition that God designed the Bible to be complete, lacking nothing, self-enclosed in truth and thus self-interpreting. Apocryphal works, writings from other contemporaneous cultures, and of course the sciences, are all treated as having little if any value in interpreting the Bible. In particular, insofar as these extra-Biblical sources are believed to be in conflict with a literal, plain reading of Scripture, they are relegated to a spiritual worldview opposed to truth.

However, there are some areas in which the Bible can be shown unequivocally to be in error based on information available outside of Scripture. For instance, Jesus referred to a mustard seed as being the smallest of all seeds on Earth when in fact smaller seeds do exist (Mark 4:31). God declares that ostriches treat their young harshly, as if they were not theirs, but in fact ostriches are extremely meticulous about how they take care of their offspring (Job 39:16). But the widely held belief that the Earth was the center of the universe and that the sun, moon, and stars rotated around it has a special place in the church's history.

Geocentrism may not have had its start in the sacred literature that became the Old Testament, but numerous passages helped to sustain this view: Psalm 19:4-6, Psalm 50:1, Psalm 104:19, Psalm 113:3, Ecclesiastes 1:5, Isaiah 45:6. Through the time of Galileo in the seventeenth century, the Church viewed geocentrism as an unambiguous truth from the Bible, so much so that in 1616 Rome issued an order to Galileo forbidding him to teach otherwise. The order expressed the position of the church in this way:

> *The proposition that the sun is the center of the world and does not move from its place is absurd and false philosophically, and formally heretical because it is expressly contrary to holy scripture. (per Santillana)*

Eventually, when Galileo was 70 years old, he was hauled before an Inquisition committee. In failing health and beset with anxiety, with failing eyesight, and being threatened with torture and imprisonment, he yielded on all counts and recanted all his anti-geocentric views. He was charged with heresy, and forced to recite a scripted statement of repentance while kneeling before the committee. The church in Rome sent copies of the final judgment to the churches in all surrounding areas, and placed limits on where Galileo could go, where he could speak, and on what topics. Humiliated, Galileo never wrote or spoke again of his studies of the movement of the sun, stars, and planets.

The church's geocentric stance in the seventeenth century is eerily similar to the modern-day literalist view of the origins narrative: it was based on a plain reading of the Bible, the science of the day was deemed to be in conflict with the Bible and thus heretical, and leaders expressed the concern that if one questions the Bible's authority and truthfulness on one point, the door has been opened to questioning its authority and truthfulness on all points. Modern literalists have been forced to distance themselves from their predecessors by distinguishing their views of the Genesis creation account as something altogether different. One way they have done this is to add a nuance to their definition of inerrancy whereby "observational descriptions of nature" are a permissible departure from a purely literal reading (per Article XIII of the 1978 Chicago Statement on Biblical Inerrancy). This allows them to admit that verses describing the movement of the sun represent only what an observer saw, not what actually happened. But literalists do not treat any scientific evidence for biological evolution or deep time as being valid extra-Biblical evidence relevant to the interpretation of the Bible because they do not consider them to be "observational;" no one can directly observe the past, and thus any suppositions about events in the past are illegitimate considerations in the interpretation of the Bible.

One of the hallmark's of the literalist approach is that the author's original intent and beliefs must be taken into account in order to correctly interpret a Bible passage's meaning. Literalists emphasize that the Genesis creation account was intended by its authors to be read as literal history, so it must be read as literal history. But the geocentric view was commonplace in the time that the Old Testament was written, such that what was written about the movement of the sun in the Bible reflected a belief that the sun was in fact moving. That the authors were recording only what they saw, not what was actually occurring, is actually irrelevant under Biblical literalism. To be true to their own approach to interpretation, literalists would need to maintain a geocentric view of the universe even in the face of modern day scientific evidence to the contrary. The fact that they are unwilling to do this demonstrates the failure of a purely literalist approach to Bible interpretation.

Prevalent imprecision

It is inappropriate to insist that every detail in the Genesis creation account must be literally true and precise when so much of the rest of the Bible contains figurative language and imprecision even in passages that are generally categorized as historical narrative. For example, Genesis 15:13 says that the Israelites would be in Egypt for 400 years, while Exodus 12:40 says that they were in Egypt for 430 years. In this particular case we have two passages that contradict one another, providing a clear indication that one (or possible both) numbers are imprecise. In a similar way, Genesis 6:3 says that henceforth God was limiting man's age to 120 years, but several hundred years later Abraham died at the age of 157 (Genesis 25:7). We cannot rule out the possibility that other passages, for which there are no parallels that could confirm imprecision, are nevertheless similarly imprecise.

In another example, the Bible seems to say something very different than it means:

Then Israel said to Joseph, I am about to die, but God will be with you and take you back to the land of your fathers. (Genesis 48:21)

It was not literally true that Joseph would return to the land of his fathers. Hundreds of years later his descendants did take Joseph's bones with them during the exodus from Egypt and the eventual entry into the promised land, but it is unlikely one would ever conclude this from Israel's statement above.

In the story of Moses and the ten plagues, we find an example of imprecise language wherein "all" actually means something less than "all." We are told that the fifth plague of disease resulted in the death of all of the Egyptian livestock:

And the next day the Lord did it: All the livestock of the Egyptians died, but not one animal belonging to the Israelites died. (Exodus 9:6)

But in preparation for the seventh plague of hail, Moses told Pharoah to bring all his livestock into shelters.

Give an order now to bring your livestock and everything you have in the field to a place of shelter, because the hail will fall on every person and animal that has not been brought in and is still out in the field, and they will die. (Exodus 9:19)

We are told that some Egyptians ignored the order from Moses and left their livestock in the fields. Those livestock were subsequently killed by hail. If all the Egyptian livestock has truly died in the fifth plague of disease, there would have been none left to die in the seventh plague of hail.

Jesus Himself spoke imprecisely on numerous occasions in order to add emphasis or to convey some point.

"Do you see all these great buildings?" replied Jesus. "Not one stone here will be left on another; every one will be thrown down." (Mark 13:2)

The temple was indeed destroyed as Jesus predicted. But there are in fact many stones that have been left atop another: the western wall in Jerusalem is the base of the temple, and it remains more or less intact to this day.

Early literalism and allegory

In support of their approach, modern day literalists say that a literal interpretation of the Genesis creation account dates back to the early church and even before. They are partially correct in that there have been theologians in every era that have been strident literalists. But there have also been many theologians who took nonliteral views of the creation account. Bouteneff notes that texts available from the second century BCE through the first century CE present neither a homogeneous view of Adam nor a consistent treatment of the creation account in Genesis Chapter 1. He goes on to say that many Greek writers in the first few centuries CE emphasized that the creation account was not about science - it did not narrate God's technology.

In the late third century, the predominant current of opinion was that the creation account should be taken literally, and by the fourth and fifth centuries a vigorous reaction against allegorism arose among theologians. But before this time, it was not uncommon for prominent theologians to propose figurative interpretations. Origen of Alexandria (3rd Century) was famous for this, though because of his unorthodox views in other areas his ideas are often sidelined. In the late 4th Century Augustine of Hippo, too, had a propensity to allegorize, and he is known to have taken a figurative view of several aspects of the creation account. For instance, he suggested that creation occurred instantaneously but was recorded as occurring in six days to accommodate the limitations of human understanding. With regard to God's command to Adam and Eve to be fruitful and multiply, Augustine said,

> *Should we understand it in a physical manner or spiritually? For we are permitted to understand it spiritually and to believe that it was changed into sexual fecundity after sin. (per Louth)*

Similarly with regard to the serpent in the garden of Eden,

> *The serpent entered the paradise spiritually and not bodily, as the apostle suggests: "You were living by the principles of this world, obeying the ruler who dominates the air, the spirit who is at work in those who rebel." (per Louth)*

The Cappadocian Fathers (4th Century) took figurative views of the creation account on several occasions. The brothers Basil of Caesarea and Gregory of Nyssa viewed Adam as equivalent to the human race rather than as an historical being (per Toews). Gregory of Nyssa refused to correlate the details of the Biblical creation account with actual events or what was then known about the laws of nature:

> *The fact of creation we accept; but we renounce a curious investigation of the way the world was framed as a matter altogether ineffable and inexplicable. (per Bouteneff)*

Their friend Gregory of Nazianzus took a noncommittal stance on the nature of the garden of Eden:

> *This creature God placed in Paradise - whatever this Paradise was! - and honored him with free choice (per Bouteneff)*

In the 8th Century, John of Damascus summed up the views of many theologians this way:

> *Some have imagined paradise to have been material, while others have imagined it to have been spiritual. (per Louth)*

In sum, to say that a literal approach to the creation account should be preferred over a figurative one because the former has historical precedent is to ignore the many alternative views dating back to the early centuries of the Christian era.

Practical problems with a literal approach

A literal view of the Genesis creation account creates a number of practical difficulties whose resolutions are either awkward or unsatisfactory. In some cases literalists devise stories to address these difficulties, minimally plausible scenarios for how something which appears highly improbable could have occurred. Such stories are fundamentally the very same type of explanation that creationists accuse evolutionists of generating to address gaps in the science. In other cases literalists merely point to the God of the gaps, a theological band-aid wherein divine intervention is used to explain anything than appears impractical or naturally impossible.

For instance, when God created all creatures, the literalist approach would have those creatures popping into existence fully formed, not only biologically, but also emotionally and psychologically. While we might imagine this occurring for creatures operating purely on instinct (insects and worms, for instance), it is considerably more difficult to imagine for creatures that operate on learned behavior such as apes, since they would have had no past wherein they might have learned those behaviors. Instead, God would have had to implant false memories into that first generation of creatures so that they would know how to forage for food and how to care for their young, and to establish a basis for group culture and hierarchies.

A manufactured history would also have had to have been created by God for every unit of space and time and matter and energy, with the result being that the universe appears to be billions of years old even though it is only a few thousand years old. Light from distant stars would be created en route from those stars to the earth, having never actually emanated from those stars but appearing as if it did. The decay rates of radioactive isotopes would be useless in attempts to date rocks and fossils since they point to a history that never existed. Continents would have been created in mid-drift, mountain ranges in mid-upheaval or deterioration, and the crater-scarred surface of the moon telling a story of an eons-long meteorite bombardment that never happened.

Before the Fall, God commanded His creatures to be fruitful and multiply. In a world without death, procreation would continue unimpeded, and in a matter of only a few years animals would completely saturate the earth. A bit of playful math tells us that, if God had created only frogs, every square inch of land would be occupied by a frog within less than ten years. The creationist solution is that God would have intervened to stop procreation before the planet was overrun. Either reproductive organs would have become nonfunctional, or all sexual desire would have ceased.

Before Eve's arrival, Adam named all the animals. This occurred entirely on day six of creation. Again, some playful math demonstrates the impracticality of this. There are around one million known species in the kingdom animalia, which would require generating ten names every second for the full 24 hours. Granted, this includes marine animals which presumably Adam was not charged with naming, but it also excludes all species that existed in the past but are now extinct, and all species that have yet to be discovered. These two additional categories could easily increase the total number of species by many more millions, making Adam's process of naming them in a single day essentially impossible.

Though not directly part of the creation account, the story of Noah's ark plays a critical role in the creationist view and thus also the literalist view. Following the landing of Noah's ark on Mount Ararat, animals would have needed to make their way to every other continent within a very short time, including those on the other side of vast oceans. The creationist solution is that either the continents moved in ways and at a speed not evident in geological record, such that there would be no oceans to cross, or that animals travelled the oceans on natural rafts of wood debris left over from the flood. Again, these are highly improbable solutions to the practical problems created by a literalist approach to the Bible, and could be solved much more easily and straightforwardly by taking a less literal approach to the Genesis creation account and admitting a prominent role for biological evolution and deep time.

Reading into the Biblical origins narrative

Finally, there are a number of elements of the Genesis creation account that are often assumed to be true despite there being no direct support for them in the text. For instance, there is nothing to indicate that Adam and Eve were created perfect or immortal, or that the world before the Fall was completely free of suffering and death. There is also nothing in the creation account that portrays a radical change in the moral state of humanity from perfect and united with God to thoroughly corrupted and separated from God. There is no indication that the sin of Adam and Eve affected the rest of creation, creating natural disasters and sickness and predation. Finally and most importantly, there is no hint of the concept of original sin as it is understood today: a corrupt nature that is passed from Adam and Eve to all their progeny, and which makes every human a sinner from birth, guilty and condemned, and incapable of seeking God. All of these are inferences drawn partially from a small number of statements in the creation account itself, and partially from certain New Testament passages, most notably in Romans Chapter 5. Many of these passages will be addressed later.

Conclusion

Given the prevalent use of nonliteral language in the Bible as a whole, one cannot conclude that the Genesis creation account must be read literally. Certainly a literal reading provides a simpler story than one which accommodates biological evolution and deep time. But simplicity is a poor indicator of truth as anyone who has studied organic chemistry knows. Moreover, a strictly literal approach to the Genesis creation account also comes with its own awkward problems that are not easily resolvable.

Nonliteral interpretations of the Genesis creation account existed in the first few centuries of the common era, though admittedly there were no attempts to accommodate biological evolution and deep time. But then again the science of the time provided no basis for these concepts just as it provided no basis for the concept that the earth orbits the sun. The Bible can and should be interpreted in light we know about the natural world, though faithful Christians must be wary of any change in understanding that treads near the core doctrines of the faith.

Chapter 2
A Perfect World

Under the Traditional Origins Paradigm, man's need for a Savior is premised on the belief that God created man perfect, and it was man's decision to disobey God that put him into a state of sinfulness from which he could not escape. In association with man's perfection, all of creation is also thought to have been created perfect; a perfect God could not have created anything less than perfect, so the logic goes. Because the theory of evolution seems so obviously imperfect, not to mention that it includes the necessary components of suffering and death, evolution would seem on its face to be diametrically opposed to a perfect creation, and thus to the traditional understanding of creation before the Fall of man. It is understandable that so many Christians have rejected evolution as a valid description of the rise of life on earth. In order to make room for evolution in Christianity, therefore, we must first address this idea that the earth was originally created in a state of perfection.

If the earth and all that was in it was initially created perfect, most would agree that it would be free from all suffering and animal death (God gives the plants as food for the animals, and plants certainty die in the process of being eaten, but since plants have no pain receptors and no consciousness, their death is typically ignored). By extension there could have been no disease and no genetic abnormalities. All other elements of the world would need to ensure that there would be no suffering and no death. Thus a perfect world is often thought of as being calm and mild, with no climatic extremes - rain but no hurricanes, wind but no tornadoes, no stifling heat waves, no frigid blizzards. Certainly there would be no droughts causing food sources to wither, and no floods to destroy your home. It would also be reasonable to assume that there were no earthquakes, no spewing volcanoes, and no meteorites smashing into the earth. There would be no carnivores, no accidents through which someone or something might get hurt, no sickness, and no aging. It would be a world in which only good, pleasant, and productive things can happen.

Scriptural basis for a perfect creation

Direct Scriptural support for a perfect creation is thin. Nowhere does Scripture say that God created earth to be perfect in the sense described above. The primary verse(s) used to directly support the idea of a perfect creation is the following:

> *And God saw that it was good. (Genesis 1:10; also verses 12, 18, 21, 25, and 31)*

Insofar as "good" is taken to mean "perfect" (which will be challenged below), these are the only verses that directly address the nature of creation before the Fall. The only other verse that might be taken broadly to be inclusive of a perfect creation is the following:

> *He is the Rock, his works are perfect, and all his ways are just.*
> *A faithful God who does no wrong, upright and just is he.*
> *(Deuteronomy 32:4)*

This verse refers to the infallible rightness of all that God does. It does not, however, provide any indication that what He does will be appealing to man, or be comfortable according to man's limited understanding. For instance, when God chooses not to heal someone who asks for healing, He is nevertheless acting according to His righteousness and perfection even though it may not appear that way to the petitioner. Similarly, one can certainly conclude from this verse that God's work in creating the earth was perfect, but it does not necessarily follow that creation was therefore free from all suffering. Suffering will be addressed in more detail in Chapter 4.

Beyond the verses above, all other verses are used inferentially. For instance, the following is used to infer that there was no predation among animals.

> *Then God said, "I give you every seed-bearing plant on the face of the whole earth and every tree that has fruit with seed*

in it. They will be yours for food. And to all the beasts of the earth and all the birds in the sky and all the creatures that move along the ground - everything that has the breath of life in it - I give every green plant for food." And it was so. (Genesis 1:29-30)

While this passage appears to say that it was God's intention that all animals as well as man should be vegetarians, it says nothing about whether death from other causes (for example, accidents, sickness, and old age) existed before the Fall. Moreover, there is at least one indication that animals were consumed by humans just after the Fall.

And Abel also brought an offering - fat portions from some of the firstborn of his flock. The LORD looked with favor on Abel and his offering (Genesis 4:4)

The only way that Abel could have used fat portions of animals as an offering is if he had killed the animals, something that apparently God approved of despite previously saying that mankind should only eat plants. More importantly, unstated is what Abel did with the remainder of the animals he killed. Abel's offering foreshadowed the sacrificial system under the later law of Moses under which the fat portions were considered to be the best part of an animal and were therefore fitting as an offering to Him. But in that system, the meat was consumed by the priests. As a proto-priest, Abel may likewise have consumed the meat from the animals he killed for an offering rather than merely throwing it out.

The only other verse used to support the idea that animal death did not exist before the Fall is an inference drawn from the curse on the serpent:

Because you have done this, cursed are you above all livestock and all wild animals! You will crawl on your belly and you will eat dust all the days of your life. (Genesis 3:14b)

"All the days of your life" is interpreted to mean that the serpent's lifetime will be limited, that this limit originated with the curse, and that the serpent is a representative of all animals. This interpretation is rather awkward in the face of the simultaneous view that the serpent is not an animal per se, but rather an animal possessed by Satan, and it is Satan that God is addressing in the curse rather than the animal that Satan possesses.

Regarding human death, it is viewed as being entirely absent from pre-Fall creation as a result of verses such as the following:

> *but you must not eat from the tree of the knowledge of good and evil, for when you eat from it you will certainly die. (Genesis 2:17)*

> *Therefore, just as sin entered the world through one man, and death through sin, and in this way death came to all people, because all sinned (Romans 5:12)*

> *For since death came through a man, the resurrection of the dead comes also through a man. (1 Corinthians 15:21)*

Human death and animal death more broadly are topics deserving a more lengthy treatment, and so are addressed separately in Chapter 5.

There are other verses that are less directly aimed at creation per se but which have been used to infer something about it. The primary ones are God's curses on Adam and Eve following their sin of disobedience.

> *To the woman he said, "I will make your pains in childbearing very severe; with painful labor you will give birth to children. (Genesis 3:16a)*

This verse has been used to suggest that there was (or would have been) no pain in childbirth prior to the Fall. It is also sometimes generalized to infer that no such thing as pain of any kind existed before the Fall.

> *To Adam he said..."Cursed is the ground because of you; through painful toil you will eat food from it all the days of your life. It will produce thorns and thistles for you, and you will eat the plants of the field. (Genesis 3:17b-18)*

This verse is traditionally interpreted to mean that thorns and thistles did not exist before the Fall, and as in other cases is sometimes broadened to mean that creation originally contained no harmful or useless plants of any kind. In some cases traditionalists generalize even further to say that the harsh and difficult world that lay ahead of Adam and Eve was entirely the result of the Fall.

Finally, some verses are drafted into service in support of a perfect creation despite having nothing do with it, such as the following:

> *The wolf and the lamb will feed together, and the lion will eat straw like the ox, and dust will be the serpent's food. They will neither harm nor destroy on all my holy mountain," says the LORD. (Isaiah 65:25)*

Some have interpreted this verse as describing not merely the attainment of a future perfect state of creation (which is its clear intention by context), but a return to a previously perfect state. Other verses that are similarly used to imply a previously perfect state of creation are any that describe God's redemptive work as "regeneration" or "renewal," such as he following:

> *he saved us, not because of righteous things we had done, but because of his mercy. He saved us through the washing of rebirth and renewal by the Holy Spirit (Titus 3:5)*

While these words can indeed mean a return to a previously perfect state, they can also mean a transformation into something new and better.

Another kind of reality

A perfect creation would not merely be a world exhibiting all the best characteristics of today's world. Instead, it would be different at the most fundamental level, a different kind of reality altogether, and not one that we can easily envision or make clear sense of.

To begin with, the absence of death, arguably the most critical element of a perfect creation in the traditional view, presents different problems depending on the cause of death. For instance, to say that animal bodies could not die due to accidents would require constant intervention by God so that, for instance, a brontosaurus could never accidentally step on a shrew and a monkey could never fall out of a tree and drown in a river. This type of divine intervention would also be needed to ensure that no suffering of any kind occurred - how else could Adam be protected from ever stubbing his toe on a rock?

With regard to natural death, we know that aging is a function of entropy (the Second Law of Thermodynamics), the tendency of all matter and energy in the universe to increase in disorder. If there were no aging before the Fall, either entropy did not exist, or once again there would have needed to be constant intervention from God to restrict or reverse the effects of entropy. It is this latter option that Stambaugh seems to envision, or which would be necessary, as he sees entropy operating without contributing to aging or physical death:

> *So, the second law was certainly functioning before the Fall. But that does not mean there was decay and physical death among the living creatures (man or sea and land animals, and birds - the nephesh chayyah) before the Fall.*

Either Dr. Stambaugh does not understand the nature of entropy, or he implicitly assumes a God of the gaps solution.

Alternatively, a world without any entropy at all would be a world that is fundamentally different from the world we know, to the point that we cannot even imagine what physical properties it would have. Matter would not be matter in the sense that we now know it. Just as difficult to imagine is how the moral sin of Adam and Even could cause the nature of the material universe to change in such a way that entropy came into existence.

Death through disease presents another set of difficulties. We might assume that inherited diseases would be precluded by perfect transcription of DNA during the process of creating egg and sperm, a process which today does occasionally produce errors leading to conditions such as Down's Syndrome or muscular dystrophy. But disease through viruses and bacteria are another matter. Bacteria and viruses are an ever-present reality in the world, and are necessary for the world as we know it to exist. Bacteria exist in our bodies to supplement our immune system and to aid in digestion. There of course would be no need to supplement our immune system in a world free from disease, though we might speculate that bacteria in our digestive system might be functional in a perfect world. Viruses, in contrast, have no known benefits except insofar as they may attack harmful bacteria. Again, the moral sin of Adam and Even would somehow have needed to cause viruses to come into existence and for some "good" bacteria to become harmful. Similarly, Adam and Even would have needed to develop white blood cells and antibodies after the Fall to ward off disease, unless we assume that they were present in their bodies when they were created despite there having been no need for them.

A world without carnivores

That all animals were originally created to be herbivores seems to be a straightforward and inarguable conclusion from Genesis Chapter 1, verses 29-30. But other passages appear to paint a different story. Consider that in God's speech to Job describing the wonders of creation, He appears to cite predation as one of those wonders:

> *Does the eagle soar at your command and build its nest on high? It dwells on a cliff and stays there at night; a rocky crag is its stronghold. From there it looks for food; its eyes detect it from afar. Its young ones feast on blood, and where the slain are, there it is. (Job 39:27-30)*

Similar passages can be found in the Psalms:

> *It was you [God] who crushed the heads of Leviathan and gave it as food to the creatures of the desert. (Psalm 74:14, brackets added)*

> *The lions roar for their prey and seek their food from God. (Psalm 104:21)*

Through some theological gymnastics it may be possible to maintain God's original intention that all animals were originally herbivores while simultaneously admitting that He sees predation as a glorious and integral part of creation. But doing so is considerably less straightforward than simply concluding that predation has always existed.

If predation was not part of God's original creation, one might reasonably infer that a vegetarian diet on the part of Jesus and His disciples would be consistent with His plan to reunite humanity with Himself, and would have been easy enough to implement. However, there is no indication that Jesus taught vegetarianism, and at least on one occasion Scripture says that He ate fish:

> *And while they still did not believe it because of joy and amazement, he asked them, "Do you have anything here to eat?" They gave him a piece of broiled fish, and he took it and ate it in their presence. (Luke 24:41-43)*

For Jesus to consume fish we must conclude that, at least at some level, Jesus considered the suffering of that fish in being caught and killed to be acceptable, and not to be avoided for the sake of inching creation back towards the state of perfection in which it was originally designed. Similarly, Isaiah's description of the blessings that God promises Israel include eating meat:

> *On this mountain the LORD Almighty will prepare a feast of rich food for all peoples, a banquet of aged wine - the best of meats and the finest of wines. (Isaiah 25:6)*

Ironically, Isaiah's prophecy of future blessings continues in verse 8 with "...He will swallow up death forever." In light of the reference to meat two verses earlier, it would seem that the death in question means human death rather than animal death. Further, insofar as this prophecy relates to a future time when fallen humanity is finally redeemed and the adverse effects of the Fall are corrected, one could logically conclude that animals would continue to die in that new creation, at least for purposes of consumption.

There are practical issues that arise if we are to believe that the original creation did not include predation. By studying modern day carnivores and herbivores, we know what physical characteristics make them distinct from one another. We know for instance the types of teeth that are designed for tearing flesh and cracking bones versus those that are designed for grinding plant material. We also know that the digestive biochemistry of carnivores and herbivores is different, such that a lion could not survive on a plant-only diet. If some herbivores turned into carnivores after the Fall, their bodies would have needed to change in significant ways, and the DNA that codes for those traits would likewise have needed to change. It is noteworthy that there are no examples in the fossil record of any modern day carnivore having a herbivorous ancestor, identical in all respects except for those traits specific to the type of food consumed.

The absence of all carnivores would presumably apply to insects and plants as well. Most spiders eat insects and use webs to do it. If all were instead herbivores, their spinnerets would have been superfluous. The spines on the front legs of the praying mantis and the venomous fangs of a tarantula would similarly be pointless. Plants that supplement their metabolism by ingesting insects, such as venus flytraps, sundews, and pitcher plants, would need to have been functionally unrecognizable compared to modern day forms and then change after the Fall, or else insects would have had to know to avoid them, a knowledge that they would then lose after the Fall.

Hints of an imperfect creation

Several Bible passages that are often assumed to describe the worldly disarray following the Fall could just as easily describe the nature of the world before the Fall, were it not for pre-conceived notions of a perfect creation drawn from other passages. For instance, the following passage suggests that it was part of the nature of the original creation to "wear out" over time.

> *In the beginning you laid the foundations of the earth, and the heavens are the work of your hands. They will perish, but you remain; they will all wear out like a garment. Like clothing you will change them and they will be discarded. (Psalm 102:25-26)*

Even in Genesis there is a hint of disorder in creation:

> *God blessed them and said to them, "Be fruitful and increase in number; fill the earth and subdue it. Rule over the fish in the sea and the birds in the sky and over every living creature that moves on the ground." (Genesis 1:28)*

What need is there to "subdue" anything unless it is in some way unruly, disordered, out of sync with the perfect ordering that God desires? Osborn goes further, saying that elsewhere in Old Testament, the word translated as subdue in Genesis 1:28 is used almost exclusively to refer to one thing: violent military conquest.

> *The implication, then, is that there is a difficult and even martial task for humanity in relation to the rest of the creation. Adam must literally "dominate," "subjugate" or "conquer" the rest of the creation.*

A number of early theologians also appeared to believe that creation was not perfect before the Fall. Irenaeus took pains to refute the idea that a perfect God can only make things that are perfect or complete from the beginning (per Cole-Turner). Cyril of Alexandria (4th - 5th Century) questioned the idea of a perfect creation that is free from entropy this way:

> *...that which may be regarded as subject to a beginning must necessarily also be subject to decay. (per Russell)*

Augustine believed that the original creation included some creatures that might be taken for carnivores:

> *We might infer that because the writer says three times "according to their kinds," our attention is called to three classes. First quadrupeds and creeping things according to their kinds...Next, the beasts are another class...that prowl about with fearsome mouths and claws...Finally, the herds make up a third class...These have no such fierce and violent ways as wild beasts, although some may attack with their horns. (per Louth)*

Beasts having "fearsome mouths and claws" and which exhibit "fierce and violent ways" are unlikely to be docile and herbivorous. Similarly, Cyril of Jerusalem (4th Century) expressed it this way:

> *Different natures of animals sprang forth from one earth at a single command - the gentle sheep and the carnivorous lion. (per Louth)*

God called His creation "good"

Among all the Scriptural support for a perfect creation, those from Genesis Chapter 1 wherein God calls His creation "good" are deemed the strongest. In the traditional view, God would not - indeed could not - have referred to His creation as being good unless it was perfect in every sense of the word.

The Hebrew word translated as "good" in Genesis is also used elsewhere in the Bible, and under the standard hermeneutic principle of allowing Scripture to interpret Scripture we can conclude that all intended meanings of the word must initially be under consideration for its use in Genesis. Unfortunately, as Whorton points out, nowhere else in Scripture is the word interpreted by Biblical scholars as meaning absolute perfection. According to Osborn, it can mean beauty, worthiness, or fitness, but never perfection. Despite this, the view among traditionalists is that the context of Genesis 1 narrows the field of possible meanings to one only, perfection. Since the context does not unambiguously indicate a state of perfection, using the context to justify this conclusion is unwarranted.

If "good" means perfection in Genesis Chapter 1, then there must be different levels of perfection: God calls His creation "good" several times before ending His work of creation by describing it as "very good." It would be as if God first thought creation was perfect, and then said that it was "very perfect." If instead "good" means functionally optimal as it was designed to be in an ordered system (per Walton), then "very good" would be an appropriate way to conclude that all the necessary pieces of the system were finally in place.

Augustine sees "very good" in a similar way as the conclusion of God's intention to make a whole out of pieces:

> *For if prudent observers consider the single works of God, they find that individually in their own species, they have praiseworthy measures, numbers, and orders. How much more then will this be true of all of them together, that is, of the universe that is filled with these individual things gathered into unity. (per Louth)*

Osborn cites one of the oldest Jewish commentaries on the creation narratives, the Genesis Rabba, compiled around the 5th Century from still more ancient sources, as saying that "very good" in Genesis Chapter 1 is a kind of wordplay that implies "good death." From this he concludes:

> *God alone is able to see all of the creation, these experts of the Torah maintained, hence God alone is also able to declare the creation "very good," its provisional sufferings and death included.*

Finally, we must be wary of ascribing to God viewpoints that are muddled with our own notions of what constitutes good and bad, as C.S. Lewis points out in *The Problem of Pain*:

> *On the one hand if God is wiser than we His judgment must differ from ours on many things, and not least on good and evil. What seems to us good may therefore not be good in His eyes, and what seems to us evil may not be evil.*

This is not to say, as Lewis caveats, that God's understanding of "goodness" is wholly different than ours, or else there would be no point in calling anything good. But that God called His creation good does not compel us to conclude that creation was originally in a state of idyllic perfection, free from all death and suffering.

We can see something of how God may have viewed His creation before the Fall in Isaiah's vision of the Lord. In that vision, Isaiah sees the seraphim flying above the throne.

> *And they were calling to one another: "Holy, holy, holy is the LORD Almighty; the whole earth is full of his glory." (Isaiah 6:3)*

Rather paradoxically, the seraphim saw the post-Fall earth as being full of the glory of the Lord despite the fact that it was also full of sin, suffering, and death. It was glorious because it was God's creation, and His glory shone through all of its imperfections. There is no reason to think that the seraphim, and God Himself, would not also have seen pre-Fall creation as being glorious, and thus "good."

The garden of Eden

If the earth were perfect, then it would seem that anywhere Adam and Eve went would be suitable. Indeed God tells them that their progeny should fill the earth, and also that they are to have dominion over all the animals of the earth. And yet, God places them into a particular place, a garden that God Himself has planted, and they are assigned the task of caring for this garden. It is apparently a special place within the world that God has created, a home within a home.

The garden of Eden may have been special because it was there that the tree of the knowledge of good and evil and the tree of life resided. Putting Adam and Eve into close proximity to these two trees was undoubtedly an integral part of God's long-term plans for the earth. But the garden also seems to be special in other ways. For instance, Adam and Eve are not merely assigned the task of caring for the garden, but guarding it as well (Genesis 2:15, Amplified version). Severian of Gabala (4th-5th Century), noting the absurdity of needing to guard against anything in a perfect world, solved the dilemma this way:

> *The text says 'work' and 'protect it.' From what? There were no thieves, travelers or people with bad intentions. 'Protect it' from what? From himself. (per Louth)*

To say that Adam's unfaithfulness to God would pollute or ruin the garden may be a reasonable solution in a perfect world paradigm, and in fact may be the only solution. But it is not an obvious reading, and a simpler one exists: the world was not, in fact, perfect, and there were both dangerous beasts and other humans on the earth that could have been a threat.

The passages discussed above suggest that the earth before the Fall may have been no different than it is today, a world where dangers existed and suffering and death were realities. If so, then the garden of Eden may have been something atypical, possibly a refuge of sorts. That God planted the garden Himself suggests that He wanted it to be especially suited for the humans who were going to reside there. One might imagine a place well-watered and having many edible plants, no dangerous animals, and plenty of raw materials from which to build shelters. There are references to the garden of Eden elsewhere in the Bible that appear to describe its particular fertility and beauty.

> *Lot looked around and saw that the whole plain of the Jordan toward Zoar was well watered, like the garden of the LORD, like the land of Egypt. (Genesis 13:10)*

> *The LORD will surely comfort Zion and will look with compassion on all her ruins; he will make her deserts like Eden, her wastelands like the garden of the LORD. Joy and gladness will be found in her, thanksgiving and the sound of singing. (Isaiah 51:3)*

If the world outside the garden of Eden was indeed less hospitable, then we may take God's curse on Adam after his disobedience as describing not a change in the world, but rather a description of the struggle he would face following expulsion from the garden. That is, the "thorns and thistles" were located outside the garden where the ground was less fertile and more difficult to cultivate. Adam would loose his life of abundant provision and ease in the garden and would be forced to work "by the sweat of your face" to obtain the food he needed. Severian of Gabala recognized the contrast between the garden of Eden and the rest of earth:

> *While Adam had been given the whole earth, he had been given paradise for his home. He could leave and go out of paradise, but there was not a habitable place for humans beyond its*

borders - only for senseless animals, four-footed animals, wild monsters and crawling bugs. (per Louth)

Conclusion

The perfect world which is assumed to have existed before the Fall under the Traditional Origins Paradigm is a world that is difficult to imagine. It is one thing to say that the climate would have been consistently comfortable and that vegetable food would have been plentiful. It is quite another thing to say that there would have been no entropy and no accidents, that every modern carnivore's teeth and digestive biochemistry would have existed in some alternative form that was oriented towards a vegetarian diet, and that there would have been no such thing as harmful bacteria or viruses. Not only do scientific investigations into the earth's past provide no evidence that such a world ever existed, but the Traditional Origins Paradigm provides no explanation for how the sin of Adam and Eve could have changed these elements of a perfect world into the imperfect world that exists today. Add to these problems the fact that the Bible itself provides almost no support for a pre-Fall creation that was perfect and includes multiple passages contradicting it, and there appears to be little reason to maintain the traditional view.

Chapter 3
The Trajectory of Creation

The primary objection to the idea that creation in its pre-Fall state was not perfect is that it appears to make God the author of suffering and death, something which is at odds with a perfectly good God whose eschatological promise is to eliminate all suffering and death. This is as much an emotional reaction as it is a doctrinal concern, and warrants a close look at God's purposes in creating the universe and the ways in which He operates within it. This will provide context for the discussion of suffering and death in the next two chapters.

Because God created the universe by an act of His will, we can confidently say that humanity exists because He wanted it to exist.

> *you created all things, and by your will they were created and have their being. (Revelation 4:11b)*

But His primary purpose in creating us was not to demonstrate His love to us. He certainly does do this, and in fact cannot act otherwise because God is love. But His primary goal in creating the universe and humanity was to glorify Himself.

> *Bring my sons from afar and my daughters from the ends of the earth - everyone who is called by my name, whom I created for my glory, whom I formed and made. (Isaiah 43:6b-7)*

> *For from him and through him and for him are all things. (Romans 11:36a)*

To some degree, the mere fact that the universe exists glorifies God.

> *The heavens declare the glory of God; the skies proclaim the work of his hands. (Psalm 19:1)*

However, God is glorified most profoundly when His sentient creatures choose to love and serve Him. This is truer still when it is difficult for them to do so, when it takes effort and an act of the will to turn from distractions, pleasures, suffering, and self to make God preeminent. One might reasonably expect, then, that God would establish the universe in such a way that His creatures must choose for or against Him, not just once but repeatedly, and in ways big and small.

God never intended the world to be static, even before the Fall of man. It was always meant to be a place of growth and development and discovery. In the broadest sense of the word, it was meant to evolve. Not in any random direction, but towards greater consistency with God's perfection. Mankind is both subject to that process and a participant in it, and there finds a multitude of opportunities to choose whom he will serve.

The God of processes

God does few things instantaneously or directly. Instead, He acts over a period of time, through processes that make use of things in the world He created. Nearly every example of God interacting with His creation in all of the Bible takes this approach, and it is one that is consistent with creation that was imperfect in its original state.

Under the creation account in Genesis Chapter 1, God creates the universe and earth over six days. He could have done it instantly, but chose to use a process instead. He then uses the dust of the earth to form Adam rather than making Adam out of nothing. Later in Genesis Chapters 6-8, God uses a flood to destroy mankind. Again, he could have simply caused all human beings to have instant heart attacks, but instead He chose to use a process that spanned many years.

Many of the miracles recorded in Scripture are, upon closer examination, the result of some type of process which was not merely a matter of God imposing Himself on the material world, but actually using the material world to accomplish His goal. Many of the plagues that Moses foretold would come upon Pharaoh were of this sort, insects and animals and disease and bad weather. The parting of the Red Sea took all night and was accomplished through a "strong east wind" (Exodus 14:21). While wandering in the desert, God punished those Israelites who complained about the journey and the lack of food and water by sending snakes to bite and poison them (Numbers 21:6). In each of these cases, God used a process that recruited the material world into His actions rather than acting more directly to address the issue by Himself.

In some cases, God takes what seems to be a circuitous route to achieve His purposes. When the king of Assyria threatened Hezekiah with the destruction of Jerusalem, God did not simply strike down the Assyrian army. Instead, God reassured Hezekiah through the prophet Isaiah with these words about the Assyrian field commander:

> *Isaiah said to them, "Tell your master, 'This is what the LORD says: Do not be afraid of what you have heard - those words with which the underlings of the king of Assyria have blasphemed me. Listen! When he hears a certain report, I will make him want to return to his own country, and there I will have him cut down with the sword.'" (2 Kings 19:6-7).*

That God would arrange the deliverance of Jerusalem in this way suggests that He has a preference for working within and through the physical and social structures of earthly life.

Even several of Jesus' miracles involved some sort of process. To heal a blind man, Jesus wiped mud made with saliva into the man's eyes and told him to wash in a nearby pool (John 9:6-7). He used a similar process to heal a deaf and mute man (Mark 7:33-35). On another occasion of healing a blind man, Jesus' first prayer appeared to only be partially effective, and a second prayer was needed for the complete healing (Mark 8:23-25). On cursing a fig tree that had no figs, it took a day for the tree to wither (Mark 11:14,20).

Processes abound in the world God created and in our relationship with Him. We do not come into being with fully formed intellects and abilities, but enter the world as little more than an assemblage of potentialities. We grow and learn, discovering what the world is about and how we fit into it. Along the way we discover that God exists and that He loves us, typically through the message brought to us by other people rather than knowledge deposited directly into us by God. Once we come into a relationship with Him, we start a process of sanctification that involves us following His guidance and developing in character and spiritual maturity. And then, too, we learn that God seems to prefer processes in meeting our needs and helping us to become more Christlike. If you pray for a new car, God may lead you to a different job. If you pray to grow in patience, God may arrange for you to wait.

Perhaps surprisingly, the development of theology is also a process. Few of the doctrines that we accept today as orthodox have existed in their current form since Pentecost. The dual nature of Christ, the nature of the Trinity, the doctrine of original sin, and even establishing the canon of Scripture occurred over centuries. God has seemed content to let His followers sort out the nature of Truth through a process involving debate, consensus, and prayer, messy and fraught with missteps though it is. We can be confident that God was leading His followers towards truth, but the path has almost never been straight or fast.

God's ways may be incontestable, but they are also inscrutable, often seeming to take longer than necessary and involve superfluous steps. Man can only speculate on His reasons in specific cases, based on the general supposition that He values processes because they make God and His creation co-participants in accomplishing His will; He works in and through His creation rather than above and outside of it. But this is exactly what one would expect for a pre-Fall creation which was imperfect. A process which takes billions of years and involves the "natural" emergence of a habitable planet, then life, and finally humanity, may seem unnecessary and possibly even crude, but it would be wholly consistent with the Biblical record of God's preference for processes. Conversely, a perfect creation resulting from the essentially instantaneous appearance of the material world would represent a nearly unique instance of such an approach for which corroborating examples in the rest of Scripture are scant.

Bringing order out of disorder

But the creation of the world was more than just a process. Aside from the first instant of the creation when matter and energy came into being, God's work was all about imposing order on that raw material, arranging it to suit His purposes, and making it productive. Some early theologians recognized that this process of bringing order out of disorder is evident in the Biblical creation narrative. For instance, Ambrose of Milan (4th Century) put it this way:

> *Scripture points out that things were first created and afterward put in order (per Louth)*

Similarly, Ephrem the Syrian (4th Century) put it into the terms of the understanding of his day, that all nature was composed of varying amounts of four basic elements, to which he adds heaven as a fifth created thing:

> *Heaven, earth, fire, wind and water were created from nothing...Therefore those five created things were created from nothing, and everything else was made from those [five] things that came to be from nothing. (per Louth)*

Ephrem evidently did not adopt the modern traditional view of each created thing popping into existence ex nihilo fully formed. Instead, raw materials were manipulated and put into order to form the structures of the universe as we know it. Clement of Alexandria (2nd - 3rd Century) likewise saw the Biblical creation account as being a description of bringing the world into a suitable state of order from some state of disorder in the distant past. In discussing God's rest on the seventh day, he said:

> *The resting is, therefore, the ordering that the order of created things should be preserved inviolate, and that each of the creatures should cease from the ancient disorder. (per Mook)*

Walton has described this process of bringing order out of disorder in his discussion of the paradox that God could call an imperfect creation "good":

> *"Good" pertained to the order that was being formed in the midst of non-order. The non-order, then, was not good, though not evil either, but the plan for continued ordering involved a process by which all non-order would eventually be resolved.*

There are also possible references in Scripture to this process of creating order out of disorder from the first instant of creation, assuming we first remove the lens of the traditional view that God's work in this regard only began after the Fall of man. To begin with, several recent theologians have suggested that the phrase "without form and void" in Genesis 1:2 can be rendered as "chaos," while others suggest it would more correctly represent "an unproductive and uninhabited place" (per Collins). But in either case, it is evident that creation was in its first form embryonic and unfit for God's creatures. It needed to be developed, cultivated, and organized. It was disorder in need of ordering.

At the other end of Scripture, Revelation paints a picture of the completion of that ordering with a look back to the former disorder:

> *He will wipe every tear from their eyes. There will be no more death or mourning or crying or pain, for the old order of things has passed away. (Revelation 21:4)*

Other translations render "old order" as "former things", "old things", or "the world of the past", while the Amplified Bible renders it more verbosely as "old conditions and the former order of things." Implied is an ancient set of conditions that have existed for as long as time itself, i.e. before the Fall of man. In this reading, God's process of creating order out of disorder includes doing away with those elements of creaturely existence that, from the beginning, have caused suffering.

One can similarly read Romans Chapter 8 as referring to God's work to overcome ancient disorder.

> *For the creation was subjected to frustration, not by its own choice, but by the will of the one who subjected it, in hope that the creation itself will be liberated from its bondage to decay and brought into the freedom and glory of the children of God. (Romans 8:20-21)*

In the traditional reading, God subjected creation to futility and corruption as a response to the Fall. But there is nothing here that demands this reading. It may, instead, be referring to God's permissive will in allowing creation to contain an element of disorder from the beginning, and of His intention to bring glory to Himself by overcoming that disorder and enlisting mankind as partners in that endeavor. There is a parallel concept in the story of Jesus healing a blind man in the Gospel of John:

> *As he went along, he saw a man blind from birth. His disciples asked him, "Rabbi, who sinned, this man or his parents, that he was born blind?" "Neither this man nor his parents sinned," said Jesus, "but this happened so that the works of God might be displayed in him." (John 9:1-3)*

Here God appears to say that the man's blindness had a mysterious purpose in that Jesus could use it to demonstrate who He was. This is little different than the statement in Romans Chapter 8 saying that the corruption of creation from ages past has a mysterious purpose in that God can use it to demonstrate His sovereignty over it.

Interestingly, the Roman's Chapter 8 passage above goes on to suggest that the frustration creation has been experiencing is a sign of the future birth of a new creation, one that is fully united with God:

> *We know that the whole creation has been groaning as in the pains of childbirth right up to the present time. (Romans 8:22)*

The reference to the pains of childbirth is meant to highlight suffering which is a necessary part of bringing forth something new; there is no reason to read this verse as in any way referencing the curse on Eve after the Fall. Thus this passage is consistent with the concept that God has been working through a process since the beginning of time to bring order out of disorder, a process which includes some level of suffering.

The trajectory of humanity

To say that God has been working since the beginning of time to bring order out of disorder may be accurate, but it is incomplete. God's work is not about abstract organization, but rather purposeful alignment of the material universe with His will. There is perhaps no better summary of God's endgame than this:

> *to be put into effect when the times reach their fulfillment - to bring unity to all things in heaven and on earth under Christ. (Ephesians 1:10)*

This uniting is not merely a matter of redeeming fallen humanity, of defeating sin and death and the devil. It includes a broader notion of aligning "all things" with His perfect will. This includes animals and insects and trees, rocks and oceans and wind. Eventually, all things will be in agreement with the plans and purposes of God.

Through the lens of the traditional view of a perfect creation, the Ephesians verse above is commonly read to mean *reunite*, not merely unite. But there is good reason to believe that the trajectory of creation towards an eschatological horizon where all things are united to God was not merely implemented after the Fall, but rather at the same instant that God began His work of creating the physical universe. If it were otherwise, if God were merely replacing what had been lost, man would return to the garden at the end of time, naked and ignorant. Instead the Bible ends with a city; the garden of Eden seems to have disappeared in the book of Revelation. Only the tree of life persists. This business of God uniting all things with Himself appears to be not so much about returning to an original state, but rather advancing to greater glory.

In the Genesis creation account it is evident that God moves from the unproductive to the productive, from sterile to organically abundant. But He quickly transitions from creating and ordering more broadly to focus more narrowly on humans. The first hint of this is when God creates the sun, moon, and stars, as the Bible explains that their purpose was not merely to provide light, but also to organize time into days, seasons, and years (Genesis 1:14). That type of organization would have only been meaningful for humans. Following the creation of man, God hands over some of the responsibility to Adam for caring for Earth and making it fully productive, giving him dominion over all creatures and the task of naming them, commissioning him to subdue the earth, and assigning him the task of tilling the garden.

From this point on, the Bible appears almost wholly concerned with humanity rather than the rest of creation, Ephesians 1:10 notwithstanding. This is not merely the result of God working to draw humanity back to Himself after the Fall, though it is certainly that. God was working with humans before the Fall to bring them to full maturity as will be discussed in more detail in Chapter 6. This fact is less evident in the Genesis creation account than in later references to it. Consider:

> *The first man was of the dust of the earth; the second man is of heaven. As was the earthly man, so are those who are of the earth; and as is the heavenly man, so also are those who are of heaven. And just as we have borne the image of the earthly man, so shall we bear the image of the heavenly man. (1 Corinthians 15:47-49)*

Here Paul is contrasting those who are like Adam with those who are like Jesus, and the progression from one to the other. But the description of Adam as being from the earth, a man made of dust, is a reference to his creation in Genesis 2:7 and thus to his pre-Fall condition. Paul is describing Adam as having been created in a natural, earthly state, a state wholly different than that of the Son of God. Adam was natural, Jesus was supernatural. Paul then says that those who become disciples of Jesus enter a process of maturation wherein they progress from those characteristics typified by Adam to those of Jesus. By submitting to God through Jesus Christ, man advances from the natural to the supernatural, yet another process.

The characteristics of the natural, earthly man are those that one might attribute to all animals: egocentrism, submission to the appetites, survival of the fittest. These lead to what for humans is sin as Paul describes in Colossians:

Put to death, therefore, whatever belongs to your earthly nature: sexual immorality, impurity, lust, evil desires and greed, which is idolatry. (Colossians 3:5)

Here Paul does not associate these sinful characteristics with post-Fall Adam, but rather with Adam's originally created nature. Man is earthly, but is called to be something else. Man has the spark of God within him and was given the advantage of intellect. He alone among the animals has both a destiny and the ability to overcome his earthly nature. But as that nature is ingrained, even instinctual, overcoming it takes effort and cooperation with the work of the Holy Spirit. The transition to Christlikeness does not come automatically nor instantly. The process is long and sometimes painful.

The process of spiritual maturation is not unlike raising a child into productive adulthood.

Folly is bound up in the heart of a child, but the rod of discipline will drive it far away. (Proverbs 22:15)

It would be a mistake to read "folly" in this verse as "sinfulness." Instead, folly is a reference to the natural desire to follow our instincts, to act egocentrically. From the viewpoint of a child, all things should be permissible - what it wants it should get. The child rails against any attempts to constrain its choices. But submission to such constraints is exactly what God intends to work into each of us. The natural way of the earthly man, as for all animals, is to follow whatever course he so desires. The supernatural ways of God are to follow a more limited set of activities, as Jesus says in Matthew:

Enter through the narrow gate. For wide is the gate and broad is the road that leads to destruction, and many enter through it. But small is the gate and narrow the road that leads to life, and only a few find it. (Matthew 7:13-14)

The trajectory of creation is bound up in the trajectory of humanity. That trajectory is not merely a transition from disorder to order writ large, but from the animalistic root nature of humanity to a humanity that reflects the nature of God. From the first moment that the spark of God entered into man's primeval ancestors when they were created in the image and likeness of God, man was set on a path towards authentic humanity, humanity as God intends it to be, a humanity in full union with God. To the degree that man cooperates with God, he moves along this path faster from animal instincts and urges to self-control, from disharmony in his relationships to harmony, from the unproductive to the productive, from the natural to the supernatural.

Conclusion

After creating space, time, matter, and energy out of nothing, God used a process to move the raw and disorganized to a state of functionality and purpose. That process was not merely natural, following physical laws, but was in some mysterious way guided by the hand of God to produce the world that He wanted to exist, a world suited to the pinnacle of His creation, mankind. At the appointed time, He then reached out to mankind to engage him in continuing the work of bringing order out of disorder not only in the natural world but also within humanity itself.

He could have done it differently. He could have made everything instantly. But as demonstrated throughout the Bible, God does almost nothing without using some sort of process that occurs over time and which uses people or the material world as His tools. One can surmise that He does this because it demonstrates His intimate connection with and involvement in His creation, not to mention His sovereignty over it. But He also does this because it serves the purpose of bringing glory to Himself through His sentient creatures. Every process involves change, and when there is change we must make choices about how to adapt while remaining faithful to God. Not only that, but God calls us to be agents of change in service to Him and towards the end goal of seeing all things united under Christ. God's ways may be mysterious, but we can be certain that He is glorified through the processes He employs.

Chapter 4
Suffering

The Traditional Origins Paradigm provides what appears to be a straightforward explanation for the existence of suffering: the disobedience of Adam and Eve brought evil into the world, and evil causes suffering. This approach is consistent with what many believe a perfectly good and loving God should be like, namely that He never wants His children to be in pain of any kind. That man does suffer is thus an indication that His perfect plan has gone awry.

Despite the simplicity and clarity of the explanation for the origin of suffering that the Traditional Origins Paradigm provides, it is widely seen as unsatisfactory because it provides no insight into suffering in any particular case. Morgan and Peterson summarize the views of traditionalists this way:

> *The biblical account of the fall helps us understand suffering in three important ways. First, we see that suffering is a consequence of sin. Second, we learn that suffering is not natural to God's good creation but is an intruder. Third, we realize that suffering contains an element of mystery.*

Morgan and Peterson admit that suffering often seems to be distributed inequitably and out of proportion to the need for moral correction, but make no attempt to resolve this dilemma. Instead they constrain their discussion to more general observations, such as the fact that suffering can be useful when operating as punishment or as discipline. In this their approach is consistent with many theologians before them. Much of their discussion is focused on Christ's sufferings on our behalf and the hope that, someday, all suffering will end.

We can indeed take comfort from Christ's sufferings as evidence of His love for us, and a future where suffering is completely and permanently eradicated is a source of hope for those who suffer. But the sufferer still struggles, sometimes mightily so. Suffering can be made worse, in fact, by a cold reference to the idea that suffering is caused by sin, either our own or the sins of others - suffering is our own fault, at least collectively. Even the best of Christians react strongly to such ideas, and as the intensity of suffering increases, so also does their demand for better answers or, more to the point, relief. One could be forgiven, then, for wondering if the traditional explanation for the existence of suffering is missing something.

Admitting our ignorance

Broadly speaking, there are two reasons that we struggle with suffering. The first concerns the direct effect that suffering has on our lives: It is distracting, vexing, sometimes overwhelming. It unmoors us and drains us. It interrupts our plans and places inconvenient burdens on our time, money, and energy. When we suffer, many other things in our lives suffer in tandem: the to-do list is ignored, leisure activities shrink, relationships become strained.

The second reason that we struggle with suffering is less obvious. It is the common Christian belief that God does not want any of His creatures to suffer. This belief amplifies the injustice of suffering by providing a contrast, a picture of what could be or will be (or should have been). This contrast limits our ability to accept suffering as in any way normal or natural. However bad suffering may be on its own, our belief that suffering has no rightful place in our lives can intensify that suffering.

To be sure, God does not want His creatures to suffer. Scripture is clear enough on that point. But this is a simplification of a considerably more complex truth. Christians have a bad habit of admitting that God is unknowable, His ways inscrutable, while simultaneously declaring that we understand certain things about Him perfectly. With regard to suffering, the Traditional Origins Paradigm has led to a certainty that God had never intended and still does not want any of His creatures to suffer. Under the Revised Origins Paradigm, suffering has a more integral and important role to play in creation.

If suffering is defined to include all forms and degrees of discomfort, physical, emotional, psychological, and spiritual, then even traditionalists would accept a natural role for suffering in at least some situations. For instance, hunger and thirst can be categorized as suffering if they are not satisfied at their earliest appearance. And surely hunger and thirst are not the results of the Fall. Hard work can cause exhaustion. The introvert can become overwhelmed by excessive interactions with others. A child suffers when compelled to eat his peas at dinnertime.

Some suffering is certainly the result of sin, directly or indirectly. Adultery can cause anguish. Greed on the part of one person or group can cause lack on the part of another. Anger can lead to violence. And of course God disciplines and punishes those who sin. But on two different occasions Jesus corrected the prevailing idea of His time that all suffering was due to sin.

> *Now there were some present at that time who told Jesus about the Galileans whose blood Pilate had mixed with their sacrifices. Jesus answered, "Do you think that these Galileans were worse sinners than all the other Galileans because they suffered this way? I tell you, no! But unless you repent, you too will all perish. Or those eighteen who died when the tower in Siloam fell on them—do you think they were more guilty than all the others living in Jerusalem? I tell you, no! But unless you repent, you too will all perish." (Luke 13:1-5)*

> *As he went along, he saw a man blind from birth. His disciples asked him, "Rabbi, who sinned, this man or his parents, that he was born blind?" "Neither this man nor his parents sinned," said Jesus, "but this happened so that the works of God might be displayed in him." (John 9:1-3)*

Even in the case of Jesus's sufferings, the Bible describes purposes that go beyond simply bearing the sins of the world.

> *In bringing many sons and daughters to glory, it was fitting that God, for whom and through whom everything exists,*

> *should make the pioneer of their salvation perfect through what he suffered. (Hebrews 2:10)*
>
> *Son though he was, he learned obedience from what he suffered (Hebrews 5:8)*

It is astonishing that Jesus, being God, would need to be made perfect and to learn how to be obedient to His Father. It is as if Jesus needed to be molded into a more perfect version of Himself, and the tool that His Father used to do the molding was suffering. As disciples of Jesus, we cannot expect to be exempt from similar treatment, because

> *The student is not above the teacher, but everyone who is fully trained will be like their teacher. (Luke 6:40)*

There are several examples of God setting up conditions for suffering only for the purpose of providing deliverance and showing His glory. We see this when childless and disgraced Elizabeth and Zechariah are visited by an angel and told that they will have a son (Luke 1:5-17). Arguably Israel's time in Egypt falls into this category.

> *Then the LORD said to him, "Know for certain that for four hundred years your descendants will be strangers in a country not their own and that they will be enslaved and mistreated there. (Genesis 15:13)*

Job is the classic example of suffering without a clear reason, and without any indication that any sin of Job's was to blame. It is notable that the author attributes Job's suffering to God despite the fact that it was Satan who directly attacked Job.

> *All his brothers and sisters and everyone who had known him before came and ate with him in his house. They comforted and consoled him over all the trouble the LORD had brought on him (Job 42:11a)*

In the account of Israel's flight from Egypt, there is a nuance to the crossing of the Red Sea that is often overlooked. At the beginning of Exodus Chapter 14, God explicitly orders Israel to turn around and position themselves next to the sea in a place where they are most vulnerable to attack from Pharaoh. Pharaoh takes the bait and goes on the offensive. The Israelites were understandably "terrified" when they see Pharaoh's army approaching. Terror is a form of suffering, in this case suffering directly resulting from God's direction. That suffering would turn to joy and relief after they crossed the Red Sea on dry land and Pharaoh's army had been destroyed, and the Israelites would glorify God in a new and more profound way than they ever had before. But for a moment, their terror had been disturbingly real and had existed precisely because God had arranged the circumstances that precipitated it.

Suffering is an inherent part of reality

The righteous suffer while the wicked live in ease. This is Job's complaint, and it is echoed in the Psalms. We see righteous suffering in the lives of the prophets and Stephen and Paul, and we see the arrogant living free from suffering in the lives of king Nebuchadnezzar and the Pharisees and Pontius Pilate. The Scriptures are almost unnecessary on this point except as validation of our own experience. In most cases, suffering seems to be uncorrelated with the need for punishment or correction, a random scattershot that touches some and not others.

The standard answer to this conundrum is that we are limited in our ability to see the purposes and plans of God, and can only trust that God is meting out suffering appropriately. This is at least partially true inasmuch as we are not omniscient. But when it comes to the suffering of the innocent, it is difficult to believe that God is decreeing suffering with some good purpose in mind; one has only to think of the baby born with spina bifida. In order to absolve God of any wrongdoing, in such cases we may place suffering under the umbrella of "living in a fallen world" where, in some unspecified way, there is a spillover effect of evil into the lives of the innocent. The result is suffering that God permits but does not decree, small comfort for those who want to believe in a God who does not want His children to suffer.

And then there are the occasions of suffering that we cannot even categorize as the result of evil, the accidents and mistakes that do not appear to be connected in any way with a world that has fallen from some previously perfect state. You accidentally crush a frog while on a nature hike. A piece of sand gets in your eye and scratches your cornea. You miss the nail and instead hit your thumb with the hammer. Few would say that the physical laws governing our material universe are themselves tainted by evil and thus unconsciously inclined to do us harm, but this would be the natural conclusion if even the accidental instances of suffering are the result of evil.

We cannot rule out the influence of the spiritual realm on the physical with regards to suffering. Demons are real and do cause suffering, as Scripture attests. But we have no basis on which to assert that every instance of suffering is the result of demonic activity. Similarly, it is also true that God does decree suffering in some cases as discipline even for righteous persons. But it is not at all clear that God decrees suffering in all cases in which He employs it towards these useful ends.

We are left to face an uncomfortable conclusion: Some, and possibly most suffering is in fact truly random, the natural result of living in a physical world which is imperfect and was always imperfect. If so, it means that much suffering does not come about through the direct action of God towards some good end, nor the activity of demons to aggravate, nor even the spillover effect of living in a fallen world. It is, instead, the result of existing. Biddle puts it this way:

> *Christianity too often effectively denies the reality of tragic suffering, wherein some combination of ignorance (absence of omniscience) and competing or mutually exclusive choices necessarily result in suffering despite all attempts to avoid it. Human beings are limited and will make mistakes. God created humanity thus. Christianity must embrace the paradox of pointless suffering.*

We see hints of this randomness in Ecclesiastes:

> *I have seen something else under the sun: The race is not to the swift or the battle to the strong, nor does food come to the wise*

or wealth to the brilliant or favor to the learned; but time and chance happen to them all. (Ecclesiastes 9:11)

That suffering is largely random does not necessarily mean that it is without meaning or purpose. God can imbue any suffering with purpose, and all the more so if the sufferer recognizes this potential. But the randomness of suffering does mean that it is in some inexplicable way essential to reality as we know it. C.S. Lewis says:

Try to exclude the possibility of suffering which the order of nature and the existence of free wills involve, and you will find that you have excluded life itself.

Bouteneff cites the 2nd Century BCE Jewish sage Ben Sira as saying that

death and suffering are not catastrophes brought on by Adam or by sin but are an integral part of God's ordering of the created world.

This is a difficult pill to swallow. We Christians are conditioned to believe that God is opposed to all suffering, though He might use it in some redemptive capacity out of necessity, and thus would never have created a world where suffering is woven into the fabric of reality. We cannot rule out the possibility that God could have created a truly perfect world that was completely free from all suffering, but as we saw in Chapter 2, it appears that He did not.

Physical pain

The face of suffering is arguably physical pain, possibly because it is often associated with visible bodily damage. This is not to diminish the reality or poignancy of other forms of suffering: emotional, psychological, and existential. Still, it is worth looking more closely at physical pain in the context of a Revised Origins Paradigm.

From a purely biological point of view, pain is an indication that something is wrong in one's body, and provides the motivation to exercise the proper care to promote healing. Pain's usefulness may not appear to be the same as saying that pain is good, but it comes close. After all, everything God created is good, and He created us with the ability to feel pain. But pain must be a special sort of good, a peculiar necessity. Without pain, we would be in a sorry state indeed, as exemplified in cases of leprosy or congenital analgesia. Individuals with such a disease or disorder experience escalating bodily damage due to the complete absence of pain, such that small injuries are ignored and evolve into significant ones. It would have been extremely dangerous for Adam and Eve to have been created without the ability to feel pain unless, as described in Chapter 2, God was constantly intervening to ensure that their bodies did not suffer even the slightest scratch.

Biologically speaking, the physical pain that humans can experience is the same as that experienced by animals, in that animals and humans share the same network of nerves that transmits messages of tissue stress or damage to the brain. But interestingly, it is animal pain, typically in the context of predation, which predominates in dialogues about the origins of the earth, humanity, and sin. We watch the jaguar killing the antelope, the hawk killing the rabbit, the octopus killing the crab, the snake killing the field mouse, the praying mantis killing the dragonfly, and we are horrified. The prey is clearly suffering, and the predator is ruthless and unsympathetic. To some it seems a given that no good God would ever create a world where predation is the norm. This view is bolstered by passages suggesting that predation is something that God intends to eliminate when Christ returns and establishes His full reign on the earth:

> *The wolf will live with the lamb, the leopard will lie down with the goat, the calf and the lion and the yearling together; and a little child will lead them. The cow will feed with the bear, their young will lie down together, and the lion will eat straw like the ox. (Isaiah 11:6-7)*

From an emotional point of view, the argument that God would not have created a world with predation is convincing. This is because we unconsciously place ourselves into the role of the prey and imagine the suffering we would experience. But C.S. Lewis argues that in doing so, we may be "...reading into the beast a self for which there is no real evidence." That is, animals may feel pain, but they do not suffer in the way that humans do. They do not consider the injustice of their predicament or wonder about the nature of the afterlife. Animals feel pain, but without a soul that pain cannot broaden into an experience of suffering as we know it. Lewis even questions the very language we use to describe the suffering of animals in the context of predation:

> *The fact that vegetable lives 'prey upon' one another and are in a state of 'ruthless' competition is of no moral importance at all. 'Life' in the biological sense has nothing to do with good and evil until sentience appears. The very words 'prey' and 'ruthless' are mere metaphors.*

Lewis' views suggest that we can draw little from predation regarding God's role in creation:

> *From the doctrine that God is good we may confidently deduce that the appearance of reckless divine cruelty in the animal kingdom is an illusion...*

This is not to say that animal pain is of no consequence to God. We can deduce from passages such as Isaiah 11:6-7 that it is not God's ideal for creation. But Jesus seemed to accept animal pain at some level since, as pointed out in Chapter 2, Jesus is recorded as eating fish, tacitly approving of the pain that fish experienced to become His meal. It is noteworthy that there are no examples of Jesus healing an animal or preventing one creature from preying on another. Although Scripture's silence on an issue is, as even Lewis points out, a poor basis on which to conclude anything about God's views on that issue, we can safely assume that if animal pain were of the same importance to God as human pain, Jesus would have lived His earthly life as a vegetarian.

The other oft-cited source of pain in dialogues about God's intentions for creation is childbirth, as it is part of the curse on Eve following the Fall.

> *To the woman he said, "I will surely multiply your pain in childbearing; in pain you shall bring forth children." (Genesis 3:16a, ESV)*

Under the Traditional Origins Paradigm, there was no pain in childbirth prior to the Fall (or would have been none had Eve given birth before the Fall). This view presumes that the significant stretching of tissues during childbirth would not have excited pain nerves prior to the Fall in the way that it does after the Fall. Either God actively blocked such nerve activity, or humans were incapable of feeling pain, both of which present their own problems as discussed earlier.

Some have suggested that this element of the curse need not refer to the creation of pain where none existed before, but could instead refer to an increase in pain from some to more. After all, one cannot "multiply" zero pain. At first glance it would appear to satisfy both sides: some form of pain exists before the Fall, and God's curse changes that pain to something dreadful. But pain in some muffled form, discomfort that is noticeable but easily ignored, would provide none of the necessary motivation to care for wounds and avoid situations that bring about painful injury; this element of the "goodness" of pain would be gone.

Others have suggested that Genesis 3:16 does not refer to physical pain at all, but rather the anguish of a mother whose children will enter a world that she has corrupted. The second half of the verse also addresses family relationships, providing some consistency in this respect to the full curse on Eve.

> *Your desire will be for your husband, and he will rule over you. (Genesis 3:16b)*

When Jesus mentions the pain of childbirth, He makes no mention of the curse, nor does He suggest that the pain is deserved. Instead, He takes a considerably more sympathetic tone.

> *A woman giving birth to a child has pain because her time has come; but when her baby is born she forgets the anguish because of her joy that a child is born into the world. (John 16:21)*

Here Jesus recognizes the reality of pain in childbirth, but associates it with a good thing: the arrival of a child. In a similar fashion, other analogous references to pain in childbirth associate it with the emergence of something good (Romans 8:22-23, Galatians 4:19). It seems at least possible, then, that the pain of childbirth which is the subject of the curse on Eve is indeed less about physical pain and more about broken relationships. One might even see in this verse Jesus downplaying any spiritual implications of pain in childbirth, as He points out that the mother forgets the pain after the child arrives; the pain ceases to have any hold on her. A curse whose effects are quickly forgotten hardly seems like an effective curse.

Suffering in heaven

In the traditional view, suffering is an exclusively earthly phenomenon, and one that will last only until Jesus returns. There has never been and will never be any form of suffering in heaven. The fact that physical healings were Jesus' most common miracle is often taken as evidence for this view - Jesus declared that "The kingdom of God has come near" and demonstrated the embodiment of that truth in Himself by removing the suffering associated with sickness, disability, and demonic oppression. He was bringing the comfort and pleasure of heaven to earth. Ipso facto, there can be no suffering in heaven.

A commonly cited passage in support of the premise that there is no suffering in heaven is this:

> *He will wipe every tear from their eyes. There will be no more death or mourning or crying or pain, for the old order of things has passed away. (Revelation 21:4)*

This passage comes at the end of Revelation; the author reports very little activity in heaven after it. Following the climax of the story when God vanquishes evil once and for all, it describes the final resolution, the end of the story. The Father has gathered His children around Him and declared that all they have suffered has finally come to an end. But if suffering ends in Chapter 21, then suffering was still very much at play before that. If taken literally, the fact that God wipes away our tears means that we must have had tears at that moment, tears indicative of some sort of suffering. Tears in heaven.

Keeping in mind that suffering is not limited to physical pain, there are indications that some suffering did exist in heaven. Consider:

> *When he opened the fifth seal, I saw under the altar the souls of those who had been slain because of the word of God and the testimony they had maintained. They called out in a loud voice, "How long, Sovereign Lord, holy and true, until you judge the inhabitants of the earth and avenge our blood?" Then each of them was given a white robe, and they were told to wait a little longer, until the full number of their fellow servants, their brothers and sisters, were killed just as they had been. (Revelation 6:9-11)*

Martyrs crying out in loud voices, demanding justice, hardly seem like the euphoric spirits we typically envision in heaven. They seem instead to be yearning for action and frustrated with the delay. The New English Bible makes their dissatisfaction more emphatic, saying that they "gave a great cry." Other translations say that the martyrs were "told to be patient a little longer," presumably because they were finding it difficult to do so. The Amplified Bible cross-references Psalm 79 where we find a similar expression of frustration with God's delay:

> *How long, LORD? Will you be angry forever? How long will your jealousy burn like fire? (Psalm 79:5)*

The martyrs under the altar were suffering. We can safely assume that they had a certainty available only in the afterlife that God was faithful and would eventually act, but the text itself suggests that such certainty was not enough to keep them in a state of contented bliss.

A bit later there is another example of suffering, this time in the form of physical pain.

> *So I went to the angel and asked him to give me the little scroll. He said to me, "Take it and eat it. It will turn your stomach sour, but in your mouth it will be as sweet as honey." I took the little scroll from the angel's hand and ate it. It tasted as sweet as honey in my mouth, but when I had eaten it, my stomach turned sour. (Revelation 10:9-10)*

The feeling of sourness in John's stomach is clearly meant to reflect the difficulty of the words he was to prophesy, but that did not make the sourness any less an actual, unpleasant sensation in his body.

While not explicit in the text, we can reasonably deduce that some form of suffering accompanied the war in heaven:

> *Then war broke out in heaven. Michael and his angels fought against the dragon, and the dragon and his angels fought back. But he was not strong enough, and they lost their place in heaven. (Revelation 12:7-8)*

We cannot hope to know exactly what occurred during this war, at least while we are on this side of the grave. But the idea of war without any suffering would strike any reasonable person as thoroughly odd. There might have been physical injuries. Certainly there would have been something like anger, distress, or grief.

The war described in Revelation Chapter 12 appears to be a culmination of the rebellion that began much earlier, apparently before the creation of man. There is no explicit description of the original rebellion in Scripture and of the battle that must have occurred then, but there are allusions to it (Isaiah 14:12-15, Luke 10:18, 2 Peter 2:4). If Michael and the righteous angels fought Satan in Revelation's final war, then they surely fought him in the original rebellion as well. And we can expect the righteous angels to have experienced something like anguish, anger, bewilderment, and sorrow when their brethren first rebelled.

God's ways are inscrutable

The Traditional Origins Paradigm asserts that the world God initially created contained no suffering of any kind, but the Bible is clear that we cannot make such a claim so confidently.

> *How unsearchable his judgments, and his paths beyond tracing out! (Romans 11:33b)*

There are several examples of God acting in ways that defy attempts at explanation and which seem to us to be unreasonable or downright unfair. Consider the fact that God allows Satan's eschatological tool the beast to not only war against Christians, but to win that war.

> *It [the beast] was given power to wage war against God's holy people and to conquer them. (Revelation 13:7a, brackets added)*

A few chapters later we see Satan's apparent defeat (Revelation 19:20), only to have God inexplicably reverse course and let Satan go on the rampage again.

> *He seized the dragon, that ancient serpent, who is the devil, or Satan, and bound him for a thousand years. He threw him into the Abyss, and locked and sealed it over him, to keep him from deceiving the nations anymore until the thousand years were ended. After that, he must be set free for a short time. (Revelation 20:2-3)*

Scripture offers no explanation, and we are left to conclude that God's actions to permit His people to suffer are part of some glorious plan whose complexities are beyond our understanding.

That God's ways are mysterious is also uncomfortably evident in Paul's description of God's sovereignty in election, critical to arguments for so-called double predestination:

> *What if God, although choosing to show his wrath and make his power known, bore with great patience the objects of his wrath - prepared for destruction? What if he did this to make the riches of his glory known to the objects of his mercy, whom he prepared in advance for glory (Romans 9:22-23)*

What is suggested here is direct and intentional action on God's part to condemn some individuals to hell from the moment they came into existence so that, by contrast, His expressions of mercy towards those destined for heaven would be made clear. The suffering of the condemned serves a broader purpose in God's kingdom beyond punishment for unrepentant sin.

Of those who are consigned to hell, one might also conclude that eternal punishment is out of proportion to the sins committed. After all, eternal punishment is essentially infinite punishment, while the sins that any mortal can commit on earth are finite. God has decided that this is appropriate, and we must accept that it must be in some inexplicable way good since God is perfectly good and all that He does is good.

A similar example of God's prerogative to fix the destiny of individuals before each is born can be found in God's differing treatment of Jacob and Esau.

> *"Was not Esau Jacob's brother?" declares the LORD. "Yet I have loved Jacob, but Esau I have hated (Malachi 1:2b - 3a)*

> *Not only that, but Rebekah's children were conceived at the same time by our father Isaac. Yet, before the twins were born or had done anything good or bad - in order that God's purpose in election might stand: not by works but by him who calls - she was told, "The older will serve the younger." Just as it is written: "Jacob I loved, but Esau I hated." (Romans 9:10-13)*

Our natural inclination is to see such actions by God as harsh and unfair. We have no choice but to let God be God, as the very next verse points out.

> *What then shall we say? Is God unjust? Not at all! For he says to Moses, "I will have mercy on whom I have mercy, and I will have compassion on whom I have compassion." (Romans 9:14-15)*

God's ways are not our ways. While we can confidently say that everything He does is good, it is inevitable that some things He does will not appear to be good to our mortal eyes. So what seems like a straightforward deduction - a good God is incapable of creating a world where suffering exists - is not so straightforward after all. We simply cannot rule out the possibility that God in fact created the world with suffering, for some good purpose that we are as yet incapable of understanding.

The suffering Christ

The most common way that Christian theologians address the problem of suffering is to point to Christ's participation in our suffering; in lieu of an answer or relief, we are to find comfort in the God who suffers with us. It is no less true under the Revised Origins Paradigm, and in fact takes on an expanded relevance.

Christ's suffering was, as highlighted earlier in Hebrews, an act on the part of the Father to perfect the Son in addition to being an aspect of the punishment for sin that He bore. But Christ's suffering was also a demonstration of God's unity with His creation. We see this fact reflected in several places.

Surely he took up our pain and bore our suffering (Isaiah 53:4)

In all their distress he too was distressed (Isaiah 63:9a)

These verses highlight Christ's identification with the human condition which is inherently beset by suffering. As discussed above, not all suffering can be attributed to sin, but Christ identifies with all our suffering, sin-spawned or not. He does so not merely because He loves us, but because the physical universe is more than just a machine that God put into motion and then stepped back from. Instead, it is His constant, ever-present activity that keeps the machine operating and, indeed, keeps the machine in existence.

If it were his intention and he withdrew his spirit and breath,
all humanity would perish together and mankind would return
to the dust. (Job 34: 14-15)

For in him we live and move and have our being. (Acts 17:28a)

Such a God cannot help but be intertwined with the activity of His creatures. As suffering existed before the Fall in the garden, so too has

God suffered with His creatures from the moment that they came into existence. Biddle puts it this way:

> *God must take responsibility for creating a world in which tragic suffering is inevitable because his creatures are limited and will make mistakes. We know that he takes responsibility because he shares in that suffering with us.*

As an inherent aspect of earthly existence, suffering is not something that could be eliminated through the cross. God did the next best thing in experiencing it with us. Jesus participates in our suffering not merely in sympathy to our fallenness, but also in oneness with our humanity. This issue is discussed in more detail in Chapter 15.

Conclusion

The Traditional Origins Paradigm asserts that the world was initially created perfect, and that prior to the Fall there was no suffering of any kind. To many, this is the only possible conclusion that can be drawn from the sacrosanct truth that a perfect and loving God can do nothing but good. In this context, it is not surprising that traditionalists reject the idea that all creatures including mankind arose from an evolutionary process fraught with suffering.

That God is perfectly good is not in question. The problem arises when man defines God's goodness based on his own limited and subjective perspective. Since we believe that uninterrupted comfort and contentment are the highest indicators of a blessed life, we assume that God also places a priority on these things and would have begun His creation with them. But God's highest priority for His creatures is not comfort but rather communion. It is from the place of communion with God that man finds his greatest fulfillment and purpose. Ultimately that communion leads to comfort and contentment, but as a consequence rather than as the goal.

God's sovereignty and inscrutability conceals much of the meaning and purpose of suffering.

> *As the heavens are higher than the earth, so are my ways higher than your ways and my thoughts than your thoughts. (Isaiah 55:9)*

This applies not only to God's activity in the world today, but also to pre-Fall creation. We cannot conclude that God would not have used an evolutionary process to bring about all life on earth simply because suffering was an inherent aspect of that process. God does not call us to understand, but rather to trust. This too is part of having true communion with Him.

Chapter 5
Death

What applies to suffering under the Traditional Origins Paradigm likewise applies to death: as there was no suffering, so also there was no death among God's creatures before the Fall. The disobedience of Adam and Eve brought upon them God's curse of death, and by extension death to all creatures. Jesus died on the cross as a proxy for us, so that the curse might be satisfied, all death would end, and humanity could be brought back into full relationship with God. Death was both the ultimate separator of man and God, and also the means through which God moved to repair that separation.

That Adam and Eve were created to be immortal is one of the first principles of the Traditional Origins Paradigm. But nowhere in the Bible is there any explicit statement to this effect, and those passages that might imply that they were immortal are ambiguous. As a result, Adam's pre-Fall immortality was not universally accepted in the first few centuries of the current era. For instance, Theophilus of Antioch (2nd Century) took a noncommittal stance on the issue, saying that Adam was neither mortal nor immortal, but rather in some neutral state prior to the Fall. Gregory of Nyssa (4th Century) was decidedly more settled in his opinion that pre-Fall Adam was mortal.

Under the Revised Origins Paradigm, death was an inherent part of reality from the first moment of creation. But death is no less repugnant to God under the Revised Origins Paradigm that it is under the Traditional Origins Paradigm, as it is an element of the disorder that God is working to eliminate. Bible passages which make clear God's ultimate goal in the matter of death are equally relevant under both paradigms.

> *On this mountain he will destroy the shroud that enfolds all peoples, the sheet that covers all nations; he will swallow up death forever. (Isaiah 25:7-8a)*

The last enemy to be destroyed is death. (1 Corinthians 15:26)

Then death and Hades were thrown into the lake of fire. (Revelation 20:14a)

The Traditional Origins Paradigm takes as its primary justification God's curse on Adam following his disobedience in Genesis Chapter 3, supplemented by several passages contrasting Adam with Christ and connecting death with sin in Romans and 1 Corinthians. But a closer look at the relevant passages reveals that they are not inconsistent with the premise that death existed prior to the Fall.

Forms of death

The Bible uses "death" to describe the end of something, but not necessarily its final end. Physical death means a complete shutdown of all bodily functions and subsequent decay of tissues, but this form of death is temporary since

there will be a resurrection of both the righteous and the wicked. (Acts 24:15b)

While the "second death" of the lake of fire described in Revelation 21:8 is permanent, it does not mean the end of existence but rather the end of all purpose and pleasure. It is the ultimate form of so-called spiritual death, the disruption of the relationship between a person and God. The second death is always preceded by spiritual death during one's earthly life, and can include disruption to other relationships as well.

For this son of mine was dead and is alive again; he was lost and is found. (Luke 15:24a)

Death can also be used to describe something as ineffectual or nonfunctional.

faith by itself, if it is not accompanied by action, is dead. (James 2:17b)

Without weakening in his faith, he [Abraham] faced the fact that his body was as good as dead - since he was about a hundred years old - and that Sarah's womb was also dead. (Romans 4:19, brackets added)

Death can also refer to cognitive disorder or emotional anguish.

The mind governed by the flesh is death, but the mind governed by the Spirit is life and peace. (Romans 8:6)

The cords of death entangled me; the torrents of destruction overwhelmed me. (Psalm 18:4)

Finally, death can refer generally to experiences marked by turmoil.

There is a way that appears to be right, but in the end it leads to death. (Proverbs 14:12)

The tongue has the power of life and death, and those who love it will eat its fruit. (Proverbs 18:21)

In some cases the Bible is clear about the type of death it is addressing, but in most cases the type of death is not explicitly stated and must be inferred. For instance, in these cases the context indicates physical death:

Indeed he was ill, and almost died. (Philippians 2:27a)

Just as people are destined to die once, and after that to face judgment (Hebrews 9:27)

In other cases the Bible appears to be speaking of spiritual death. For instance, Jesus says that the saved will never see death despite the fact that everyone physically dies.

Very truly I tell you, whoever obeys my word will never see death. (John 8:51)

The most logical conclusion is that He is speaking of spiritual death, not physical death. Similarly, in the passage below Paul says that sin leads to death, but he then contrasts death with righteousness rather than with physical life. Again, the logical conclusion is that the death of which he speaks is spiritual death.

> *Don't you know that when you offer yourselves to someone as obedient slaves, you are slaves of the one you obey - whether you are slaves to sin, which leads to death, or to obedience, which leads to righteousness? (Romans 6:16)*

Jesus Himself used the word "death" in both ways, sometimes in the same sentence.

> *Jesus said to her, "I am the resurrection and the life. The one who believes in me will live, even though they [physically] die; and whoever lives by believing in me will never [spiritually] die. (John 11:25-26a, brackets added)*

In some cases it is not exactly clear what form of death is being described. For instance, in the passage below Jesus refers to death twice, and His phrasing suggests that He intended the word death to have the same meaning in both instances.

> *I am the bread of life. Your ancestors ate the manna in the wilderness, yet they died. But here is the bread that comes down from heaven, which anyone may eat and not die. (John 6:48-50)*

But if Jesus was using death consistently, His point is obscure. Either He meant that the Israelites spiritually died in the wilderness, i.e. were eternally separated from God, or He meant that by eating of the bread of life one would not physically die. The original audience would have reasonably assumed that Jesus was offering freedom from physical death through eating His own flesh, in opposition to the Israelites who physically died in the desert despite eating the manna provided by God. Only later would they learn that He was offering freedom from spiritual death, not physical death. As discussed in more detail in Chapter 14, it is possible that Jesus saw physical death followed by resurrection to eternal life as being equivalent to escaping physical death altogether, but this is not evident in the passage above. We are left to wonder if Jesus was intentionally conflating physical and spiritual death, and had expected only His true followers to unravel the meaning of this passage through further inquiry as He intended with many of His parables.

Another verse where the type of death is unclear is the following:

For the wages of sin is death, but the gift of God is eternal life in Christ Jesus our Lord. (Romans 6:23)

While traditionists often see death in this verse as meaning physical death, Paul contrasts it with eternal life, i.e. joyful communion with Jesus that lasts for all eternity. This is primarily a spiritual state, though it is supplemented by physical resurrection. The most straightforward reading, though not the only possible reading, is that Paul was here referring to spiritual death rather than physical death.

The curse of death

The fact that death is used in a variety of ways in the Bible, with meanings that are not always self-evident, is important for the interpretation of its use in the context of the story of Adam and Eve.

God gives a very clear warning to Adam about the consequences of eating fruit from the tree of the knowledge of good and evil.

And the LORD God commanded the man, "You are free to eat from any tree in the garden; but you must not eat from the tree of the knowledge of good and evil, for when you eat from it you will certainly die." (Genesis 2:16-17)

Under the Traditional Origins Paradigm, the primary referent in this verse is physical death in keeping with one of its core premises that Adam was created immortal. But the fact that Adam did not physically die on the very day that he ate from the tree of knowledge has by necessity led traditionalists to both reinterpret the meaning of "day" as well as to supplement the referent to include spiritual death.

If there was no such thing as physical death prior to the Fall, Adam would have had no examples of death in his experience to enable him to understand the gravity of God's warning. Traditionalists solve this dilemma by positing some special knowledge about the nature of physical death that was imparted to Adam at the time of his creation, in the same way that Adam must have been given the ability to use language and take care of the garden without having had to learn these skills. Under a Revised Origins Paradigm, in contrast, Adam understood the concept of physical death because he lived in a world where it was the expected end for all creatures, humans included. God's warning would then have meant that eating from the tree of knowledge would lead to premature death, almost as if the tree's fruit was poisonous (which in some supernatural sense it most certainly was).

More importantly, God's warning to Adam very likely did mean something other than or in addition to physical death. The actual curses given to the serpent, Eve, and Adam in Genesis Chapter 3 are considerably more detailed and more broadly applicable than God's original warning to Adam which said nothing of suffering, enmity, discordant relationships, or toilsome labor. Even so, God deemed His simple warning to be sufficient. It may be, then, that in His initial warning to Adam He was using death as a shorthand for all of the curses. In this case, "death" in Genesis 2:17 would not have meant only physical death or even spiritual death, but rather the comprehensive end of all the special blessings and protection He had planned for Adam and Eve.

Following Adam's disobedience, the only portion of God's curse relating to physical death is the following:

> *By the sweat of your brow you will eat your food until you return to the ground, since from it you were taken; for dust you are and to dust you will return. (Genesis 3:19)*

While it is clear that Adam will physically die, it is not clear that he will die because of his sin. Instead, God says that Adam will return to the ground because ("for") he was taken out of the ground. His body was built from the raw materials of the Earth, and his destiny is for those raw materials to disaggregate and return to their original, lifeless state. Everything in God's phrasing suggests a natural cycle rather than a curse per se, though it is still that as discussed later. Toews draws essentially the same conclusion:

> *Death is not meant as a punishment; it is referred to as part of the natural order of things.*

Athanasius, Bishop of Alexandria in the 4th Century, similarly suggested that death was the natural destiny of Adam and Eve:

> *But if they [Adam and Eve] transgressed and turned back and became evil, they might know that they were incurring that corruption in death that was theirs by nature. (per Louth, brackets added)*

We see hints elsewhere in Scripture that physical death was always in Adam's future, regardless of whether or not he was disobedient. Ecclesiastes uses almost exactly the same phrasing as Genesis, again suggesting a natural cycle:

> *All go to the same place; all come from dust, and to dust all return. (Ecclesiastes 3:20)*

The Psalmist suggests that physical death was part of God's purpose in creating humanity:

> *Remember how fleeting is my life. For what futility you have created all humanity! Who can live and not see death, or who can escape the power of the grave? (Psalm 89:47-48)*

God created humanity for "futility," which in the following sentence is associated with death. It is therefore possible that the Old Testament at least supports the concept of death as natural and thus existing prior to Adam's sin.

Even so, several passages in the New Testament, most notably in 1 Corinthians and Romans, seem to more directly connect Adam's physical death with his disobedience, notwithstanding the fact that the "death" of which they speak is usually ambiguous with regard to being physical versus spiritual. In order to properly address those passages, it is helpful to first review the role of the tree of life in the garden of Eden.

The tree of life

God's punishments for Adam and Eve are immediately followed by their banishment from the garden. Traditionalists view the banishment as another element of the overall curse, but the Bible explicitly tells us that the driving factor was not the curse per se, but rather the potential for Adam and Eve to eat from the tree of life and thus live forever.

> *And the LORD God said, "The man has now become like one of us, knowing good and evil. He must not be allowed to reach out his hand and take also from the tree of life and eat, and live forever." So the LORD God banished him from the Garden of Eden to work the ground from which he had been taken. (Genesis 3:22-23)*

This is the only place in Genesis that describes the purpose of the tree of life. The only other explanation of its purpose is provided in Revelation.

> *On each side of the river stood the tree of life, bearing twelve crops of fruit, yielding its fruit every month. And the leaves of the tree are for the healing of the nations. (Revelation 22:2b)*

In addition to healing and immortality, other passages that make analogies to the tree of life provide other possible purposes:

Wisdom:	*She [wisdom] is a tree of life to those who take hold of her; those who hold her fast will be blessed. (Proverbs 3:18, brackets added)*
Relief:	*Hope deferred makes the heart sick, but a longing fulfilled is a tree of life. (Proverbs 13:12)*
Consolation:	*The soothing tongue is a tree of life, but a perverse tongue crushes the spirit. (Proverbs 15:4)*

One might justifiably conclude that the tree of life is a source (or the source) of all things which make life good in all ways, physical, psychological, and emotional. But if so, what need was there for it to be in the garden of Eden? According to the Traditional Origins Paradigm, every aspect of existence before the Fall was perfect. There would have been no need for Adam and Eve to partake of its fruit as they were already immortal, happy, and fulfilled in every way. That the tree could have restored to Adam and Eve what they lost through their disobedience is irrelevant to the question of why God placed the tree in the garden in the first place inasmuch as God prevented their access to it following the Fall. Under the Traditional Origins Paradigm, the presence of the tree of life in the garden was essentially pointless.

Genesis gives no indication one way or the other regarding whether Adam and Eve knew of the tree of life or ate from it prior to the Fall. All that is clear is that if they had eaten of it after the Fall, they would have lived forever. Was one bite from the tree (fruit or leaves, Scripture is ambiguous) all it would have taken? According to the description of the tree of life in Revelation, it yields its fruit "each month," implying that those with access to it are eating from it on a regular basis. Thus it is plausible that procuring all the benefits that the tree offered would require that one eat from it on a regular basis.

If Adam and Eve were ordinary humans, chosen by God out of the many who existed on earth to participate in something new, their natural lifespans would likewise have been ordinary. But if they had begun eating from the tree of life during their time in the garden prior to the Fall, the aging process would have been temporarily interrupted. Their bodies would have been given a new level of vigor not unlike that given to Isaiah after his confrontation with the prophets of Jezebel:

> *Then he lay down under the bush and fell asleep. All at once an angel touched him and said, "Get up and eat." He looked around, and there by his head was some bread baked over hot coals, and a jar of water. He ate and drank and then lay down again. The angel of the LORD came back a second time and touched him and said, "Get up and eat, for the journey is too much for you." So he got up and ate and drank. Strengthened by that food, he traveled forty days and forty nights until he reached Horeb, the mountain of God. (1 Kings 19:5-8a)*

The effects of the food and drink offered to Isaiah by the angel were temporary, but they nevertheless gave him a supernatural strength. It would not be inconsistent for the tree of life to have operated in the same way for Adam and Eve prior to the Fall: by eating of it, they would have experienced a temporary, supernatural boost to their physical bodies that would have included an interruption of the normal aging process. Immortality would have been achieved so long as they continued eating from the tree of life indefinitely. Ambrosiaster in the late 4th Century described nearly exactly this (per Kelly), and his contemporary Augustine described something similar in the context of his musings about what life would have been like had there been no Fall.

> *Rather those who had begotten children would remain in the prime of life and would maintain their physical strength from the tree of life that had been planted in paradise. (per Louth)*

Although Adam and Eve no longer had access to the tree of life after being banished from the garden, the benefits of having previously eaten from the tree of life would undoubtedly have lingered in the same way that they did for Isaiah. It is thus possible that their lifespans had increased dramatically. We might even speculate that the physical benefits of having eaten from the tree of life were also passed on to their children, resulting in extended lifespans for them as well. But the promise of immortality would be gone, lost immediately upon their banishment from the garden.

Paul on Adam and death

If the purpose of the tree of life was to confer immortality to mortals, then the passages in Romans Chapter 5 and 1 Corinthians Chapter 15 which make a connection between Adam's sin and death for all take on a new meaning. These passages are of premier importance for both the Traditional and Revised Origins Paradigms and will be discussed in more detail in Part III in the context of original sin. Here we focus only on those portions that are most directly related to Adam's death.

Therefore, just as sin entered the world through one man, and death through sin, and in this way death came to all people, because all sinned (Romans 5:12)

many died by the trespass of the one man (Romans 5:15b)

The judgment followed one sin and brought condemnation (Romans 5:16b)

by the trespass of the one man, death reigned through that one man (Romans 5:17a)

one trespass resulted in condemnation for all people (Romans 5:18a)

For since death came through a man, the resurrection of the dead comes also through a man. For as in Adam all die, so in Christ all will be made alive. (1 Corinthians 15:21-22)

These passages appear to make two main points regarding death:

- Death appeared in the world for the first time as a direct result of Adam's sin
- Adam's sin resulted in all humans being condemned to death

These two points remain true under the Revised Origins Paradigm, but their precise meaning is somewhat different than under the Traditional Origins Paradigm.

As described earlier, the passages which correlate death with Adam's sin may not be speaking of physical death at all. There is nothing to prevent us from seeing Adam's sin as resulting in only spiritual death: a break in the relationship between he and God and the beginning of a pattern of acquiescing to the disobedient inclinations of his soul. Spiritual death would also be more consistent with the fact that the act which led to punishment - seeking forbidden knowledge - was an act that affected Adam's soul and/or spirit, rather that his body.

Nevertheless, it is appropriate to consider how Paul's link between Adam's sin and the entrance of death into the world could be true under a Revised Origins Paradigm if the death to which Paul refers was indeed physical. To this end, we must consider the broader plan that God had in calling Adam and Eve into the garden in the first place.

God's direct and frequent interaction with Adam and Eve, as well as the unprecedented command given to them and the associated consequences of disobedience, appears to have been the inauguration of an entirely new stage in the relationship between God and humanity. As described in more detail in Chapter 7, God's intention was to develop a new community of humans devoted to Him and to use that community to influence the rest of the world. God's purposes for Adam and Eve would be repeated many times in the rest of the Bible: reaching out to a single person or small group, training them into a more devoted and intimate relationship with Himself, and then sending them out to others to carry the message of God's love and faithfulness, and of the need for our fidelity to Him in order to secure the blessings that He offers.

However, Adam and Eve's calling was different from all subsequent callings in a very important way: they had access to the tree of life. Their destiny was not merely to be examples to the rest of the world of what it means to know God and be faithful to Him. They were intended by God to also demonstrate the benefits of that relationship with God through their vigorous physical health, their exuberant optimism, and their robust sense of meaning and purpose. And of course their immortality which was to be obtained through eating from the tree of life. Those to whom God sent Adam and Eve would have the opportunity to come into a new relationship with the God that Adam and Eve served, after which they too would have the same access to the tree of life. Thus the final outcome of God's calling to Adam and Eve was to be that all of humanity would become immortal. Death was not humanity's true destiny.

But Adam and Eve's disobedience interrupted God's plans. The immortality that was theirs for the taking slipped through their fingers, and with it the hope of immortality for all humans. Death became a permanent fixture, and thus "reigned" in a way that it had not before. Insofar as Paul was referencing physical death in his discourse on Adam's sin in Romans Chapter 5 and 1 Corinthians Chapter 15, we might see death as becoming humanity's destiny through Adam despite the fact that physical death had existed outside the garden and before the Fall.

We must also consider what it meant for death to come "into the world" per Romans 5:12. The traditional reading would make this phrase applicable to the entire earth, but in other contexts references to the world clearly indicate something more limited.

> *What good is it for someone to gain the whole world, yet forfeit their soul? (Mark 8:36)*

> *In those days a decree went out from Caesar Augustus that all the world should be registered. (Luke 2:1, ESV)*

> *So the Pharisees said to one another, "See, this is getting us nowhere. Look how the whole world has gone after him!" (John 12:19)*

When death entered the world, it may have done so in a geographic area where it did not exist previously, or was not meant to exist. The logical applicable area would be the garden of Eden, the "world" where Adam and Eve lived in communion with God. It was there that the promise of immortality existed, and it was there that Adam's disobedience shattered that promise. Physical death did not appear for the first time on earth when Adam sinned, but it did appear where it should not be. Death was an invader in the very space the God had set aside to be free from death.

After God banished Adam and Eve from the garden of Eden, He ensured that they would never again have access to the tree of life.

> *After he drove the man out, he placed on the east side of the Garden of Eden cherubim and a flaming sword flashing back and forth to guard the way to the tree of life. (Genesis 3:24)*

If the way to the tree of life was blocked for Adam and Eve, it was also blocked for all of their descendants and for every other person on the earth. In a very real sense, Adam's sin foreclosed the possibility of immortality for all people. All humans were condemned to physical death without hope of escape because Adam's choice made it so.

The continuation of death after the cross

Jesus' sacrifice on the cross appeased the wrath of God towards sin, thereby removing the condemnation that rests on all those who place their faith in Him. That condemnation included a sentence of death for every individual. The Bible is clear that Christ's work was meant to address the punishment of death.

> *our Savior, Christ Jesus, who has destroyed death and has brought life and immortality to light through the gospel. (2 Timothy 1:10b)*

In place of "destroy" other translations use "abolish". As discussed above, Jesus Himself said that faith in Him would mean freedom from death. But His words often blurred the distinction between physical death and spiritual death, and His audience could be forgiven if they thought that He was offering immortality, a miraculous escape from physical death. It is only through the rest of the story, namely Jesus' death and the resurrection, that we know what Jesus really meant. Paul tells us plainly that the remedy wrought by Jesus is not physical immortality but rather resurrection to eternal life with God.

> *For if we have been united with him in a death like his, we will certainly also be united with him in a resurrection like his. (Romans 6:5)*

It should be enough for us to know that Jesus made a way for us to be freed from the condemnation brought about by sin, to have peace with God, and to live eternally with Him. That this way is through resurrection after we die rather than by avoiding physical death altogether should ultimately be of no consequence. Nevertheless, one might reasonably ask, why do we still physically die?

Under the Traditional Origins Paradigm, God's punishment on Adam for his sin included physical death. At the same time, the Bible is clear that Jesus took upon Himself the punishment for man's sins, including death.

But he was pierced for our transgressions, he was crushed for our iniquities; the punishment that brought us peace was on him, and by his wounds we are healed. (Isaiah 53:5)

...by the grace of God he might taste death for everyone. (Hebrews 2:9b)

John Calvin describes the complete transfer of our guilt and the associated punishment to Jesus this way:

Our acquittal lies in this: that the guilt which made us liable to punishment was transferred to the head of the son of God...This had to be done so that the whole curse, which we deserved because of sin, might be taken from us by being transferred to him. (per Lane)

If the condemnation and punishment for sin that applies to all people was entirely removed by the cross, and physical death was part of that punishment, then physical death should likewise have been abolished by the cross.

Therefore, there is now no condemnation for those who are in Christ Jesus, because through Christ Jesus the law of the Spirit who gives life has set you free from the law of sin and death. (Romans 8:1-2)

And yet, this is not the case. We all still physically die despite no longer being under the curse of Adam.

There are several possible ways to resolve this dilemma under the Traditional Origins Paradigm. One might, for instance, conclude that the cross does not address the punishment of physical death at all, but rather only spiritual death. Our relationship with God is repaired and we escape the second death, eternal separation from God in hell, but the cross has no bearing on the punishment of physical death. But under the Traditional Origins Paradigm the original dual punishment of physical and spiritual death given to Adam comes as a package deal, and there is no obvious reason that the cross would address one and not the other.

An alternative approach which straddles the fence between the Traditional and Revised Origins Paradigms is the view that despite Jesus bearing our sins on the cross, we nevertheless retain sin in our lives, and this sin results in physical death. In this approach, the punishment for sin is removed, but the presence and natural consequences of that sin are not. Physical death is one of those natural consequences, not a punishment imposed on man because of sin. The Bible recognizes the ongoing reality of sin in the lives of believers even though the punishment for that sin has been removed.

> *He himself bore our sins in his body on the cross, so that we might die to sins and live for righteousness; by his wounds you have been healed. (1 Peter 2:24)*

The healing to which this verse refers must be spiritual death only. Both the activity of sin and physical death remain, and it may be reasonable to assume a causal connection between the two.

Ultimately, the Revised Origins Paradigm provides a more straightforward explanation for why we physically die despite the fact that Jesus bears all of God's punishments on our behalf: we die because physical death is not a punishment, and neither is it a natural consequence of sin. Instead, every human being that has ever existed was always destined to die by virtue of having a natural body generated through an evolutionary process, and the one opportunity to gain immortality was lost through Adam's disobedience. While the cross aims to restore fallen man in addition to redeeming him, physical immortality cannot be restored because man was never immortal to begin with.

Death as mercy

Notwithstanding the discussion above, we may be justified in viewing physical death as an invader into the ideal world that God had planned for His children. Collins describes the common view of death this way:

We can accept the death of our bodies, we can acquire the spiritual skills to die well; but we still see death as an enemy. Hence we have grounds for thinking that bodily death was never God's intended outcome for his human creatures.

Even so, some early theologians cast death in a more positive light, death as mercy. God knew that Adam and Eve's future outside of the garden was going to be difficult and in His love, which was not diluted by their rebellion, He used death to limit their earthly suffering. Ephrem the Syrian said that God implemented death

...lest Adam and Eve, after having eaten of this tree [of knowledge], live forever and remain in eternal lives of suffering... (per Louth, brackets added)

Theodoret of Cyrus in the 5th Century similarly saw death as mercy.

For death brings about separation from this life and brings evil works to an end. It sets us free from labor, sweat and pain, and ends the suffering of the body. Thus the Judge mixes his love for us with punishment. (per Louth)

Death may also be a means through which God draws wayward humans back to Himself. Adam and Eve failed to trust God regarding the dangers of the tree of knowledge, and that failure led to, among other things, physical death. Death stood as a finality that they could not overcome or escape by any exertion of will or might. Their only recourse was through seeking and submitting to the God that they had mistrusted and disobeyed.

Impending death also creates a fear, which may or may not be conscious, of the annihilation of self or of any meaningful existence. This fear can compel some to pursue excessive pleasure as a way of drowning out that fear.

If the dead are not raised, "Let us eat and drink, for tomorrow we die." (1 Corinthians 15:32b)

Others pursue activities that are unhealthy or self destructive, such as death-defying, adrenaline-producing stunts in an attempt to prove their mastery over death. Fear itself or the habits to which it leads are forms of slavery, and only a renewed trust in and submission to God can free us from that slavery.

> *Since the children have flesh and blood, he too shared in their humanity so that by his death he might break the power of him who holds the power of death - that is, the devil - and free those who all their lives were held in slavery by their fear of death. (Hebrews 2:14-15)*

Death is thus a two-edged sword, both the result of the disobedience of Adam and Eve and a means through which God woos mankind back to Himself.

Conclusion

The traditional view of death is that it is an invader in God's originally perfect creation, entering human experience as the direct result of the disobedience of Adam and Eve. But the existence of the tree of life suggests that Adam and Eve were originally mortal. Death would have been repressed indefinitely so long as they regularly ate from the tree of life, but the hope of immortality ended after their disobedience when their access to the tree of life was cut off.

Moreover, the death of which God spoke in warning Adam against eating from the tree of knowledge was not restricted to physical death as evidenced by the range of punishments that He actually decreed after the Fall: Adam and Eve would not merely die physically, but life would become more difficult and more disordered. In essence, the spirit of death entered their lives in multiple ways. In subsequent generations, death would manifest not only in stress and striving, but also in greater sinfulness and eventually a near complete disconnection from God.

In God's compassion for man's helpless condition, He sent Jesus to deliver mankind from the curse of death by taking death upon Himself. In doing so He took on death in its broadest sense, all the disconnection and disorder that characterizes what we think of as normal life. Through His resurrection He provided the antidote to physical death, and in the age to come,

There will be no more death (Revelation 21:4a)

Part II

Adam and Eve

Chapter 6

The Creation of Adam and Eve

The Revised Origins Paradigm admits that Adam and Eve were real people, but asserts that they were neither the only people on earth at the time of the events in the garden of Eden nor the first to appear on earth. This paradigm requires an interpretation of the creation story in Genesis Chapters 1 and 2 that deviates from the traditional one. But this is not new. Many theologians have done this, with some going further than is necessary to the point of making Adam and Eve out to be allegories for all humans or even making the full story to be little more than a Christian myth. For the Revised Origins Paradigm, it is only necessary to determine how it could be that Adam and Eve were not the only humans on earth while remaining the progenitors of sin and death for all. This leaves a core element of the creation story intact while accommodating the scientific evidence about biological evolution and human origins.

To this end, this chapter addresses the creation of humanity in two phases encompassing two different ways of becoming, consistent with God's predilection for processes as described in Chapter 3. The first phase is the appearance of creatures on earth that can be called human. That which was not human (ancestral apes or their hominid descendants) became that which is human over countless generations, at some point taking on the image and likeness of God. It is this phase where talk of the overlap (or lack thereof) between biological evolution and the Biblical creation narrative is often centered.

The second phase is the appearance of Adam and Eve, two particular individuals among many whom God led into a garden within the region of Eden. These two would become the first of God's appointed emissaries and the means through which God intended to bring all humans on earth into a new relationship with Himself. The story of God's relationship with humanity, or at least that part of the story that is relevant to redemption, begins with Adam and Eve.

Just two of many

There are two hints in the Bible that Adam and Eve were just two of many at the time of their "creation." Much has been made of the fact that the Hebrew word *adam* means mankind. While it is clear that Adam is treated as a singular person in Genesis Chapter 2, this treatment need not apply in Genesis Chapter 1 where it is general creation which is being described. Moreover, Genesis Chapter 5 actually contains a recapitulation of the Chapter 1 account of the creation of man that makes this evident:

> *He created them male and female and blessed them. And he named them "Mankind" [adam] when they were created. (Genesis 5:2, brackets added)*

Notably, both the King James and Amplified translations actually use "Adam" here, capitalized as a proper name but applied to a plurality of humans. Admittedly the plural could refer to Adam and Eve, but there is nothing preventing it from referring to a much larger group of humans. Walton puts it this way:

> *... the Adam and Eve account in Genesis 2 could have come after an en masse creation of humanity in Genesis 1 ... though Adam and Eve should be considered as having been included in that group.*

The other Biblical passage suggesting that Adam and Eve were not the only humans on earth can be found in association with the story of Cain killing his brother Abel in Genesis Chapter 4. When God confronts and then condemns Cain, he responds with this:

> *Today you are driving me from the land, and I will be hidden from your presence; I will be a restless wanderer on the earth, and whoever finds me will kill me. (Genesis 4:14)*

Under the Traditional Origins Paradigm, the "whoever" to whom Cain refers must be members of his own family. Since at the time of Abel's death the only other humans that the Bible explicitly mentions are Adam and Eve, these are the only two that could have threatened Cain. Traditionalists also contend that no one other than Cain's own family would have a reason to kill Cain as an act of vengeance for murdering Abel. But there is nothing in the fear that Cain expresses to indicate that he was concerned about vengeance from his family particularly. He may instead have been concerned about the plight of a single, defenseless person wandering among people of other cultures and languages. Moreover, Cain calls himself a fugitive in a number of other translations, likely meaning that he is fleeing from his parents. The threat he fears, then, would have had to come from people other than his parents since he will no longer be in the vicinity of his parents.

Another possibility that traditionalists have suggested is that Adam and Eve had other unidentified children at around the same time that they had Cain and Abel, and these others had already grown to adulthood, moved away and begun settling in other areas. At the time that Cain killed Abel, his own siblings would have been living elsewhere and Cain would have encountered them after leaving his parents. A variant of this idea is that a considerable amount of time had passed between Cain murdering Abel and God's curse on Cain, on the order of decades or centuries. In that timeframe, we know that Adam had other children.

> *When Adam had lived 130 years, he had a son in his own likeness, in his own image; and he named him Seth. After Seth was born, Adam lived 800 years and had other sons and daughters. (Genesis 5:3-4)*

Again, these siblings of Cain could have grown to adulthood and spread out to settle other areas.

It would require less speculation to treat the "whoever" to whom Cain refers as other peoples who already lived in surrounding areas and who were not related to his family. This approach would also more easily explain how Cain could have found a wife in an area far from Eden, though it is not impossible that his wife was actually a sibling or some other relative as traditionalists contend. Ephrem the Syrian recognized that Cain most likely found a wife among peoples to whom he was not related:

> *Cain, therefore, separated himself from his parents and his kin because he saw that they would not intermarry with him. (per Louth)*

Whatever ambiguity there may be in Genesis about the existence of other humans at the time of Adam and Eve's "creation," two New Testament references to Adam suggest that Paul believed Adam to have been the first human being on earth. In explaining to the Athenians how the true God differed from their idol gods, he said:

> *From one man he made all the nations, that they should inhabit the whole earth; and he marked out their appointed times in history and the boundaries of their lands. (Acts 17:26)*

The Athenians had no concept of a single God that was responsible for the creation of everything and everyone, and in fact thought themselves special in regard to their origin, believing that their ancestors came directly from the soil under Athens. In order to introduce them to the true Creator God, Paul had to explain that all men had the same origin. In this context Paul does appear to say that all humans were descended from "one man." Traditionalists contend that this man must be Adam, but in fact a stronger case can be made that Paul is instead referring to Noah. The only place in the Old Testament where the origin of nations can be found is in Genesis Chapter 10 where all the descendants of Noah's three sons are listed. The chapter ends with the following summary:

> *These are the clans of Noah's sons, according to their lines of descent, within their nations. From these the nations spread out over the earth after the flood. (Genesis 10:32)*

The language is similar to that used by Paul, suggesting that Paul had this very text in mind as he spoke to the Athenians.

The other Pauline passage suggesting that Adam was the first human on earth is the following:

So it is written: "The first man Adam became a living being" ; the last Adam, a life-giving spirit. (1 Corinthians 15:45)

The Old Testament text that comes closest to that which Paul appears to be quoting is Genesis 2:7. Given the introductory phrase "So it is written," we would expect to find an exact quote, at least insofar as a translation from Hebrew to Greek allows. Instead we find that Paul has modified the text he is quoting in a distinctive way:

Genesis 2:7: the man became a living being

1 Corinthians 15:45: The first man Adam became a living being

If Paul's focus in 1 Corinthians Chapter 15 had been a theology of creation, we might take his modification of the Genesis text as a Spirit-borne revelation of the fact that Adam was the first human being to have existed on earth. But his focus is not the creation story in 1 Corinthians Chapter 15. His focus is on Jesus. The "first man Adam" is Paul's way of identifying where the story of God's (essential) interaction with humans begins and where things went wrong, in order to provide a contrast to the culmination of God's efforts to set things aright through Jesus. Just as Jesus is not the last human, neither must we read Paul's paraphrase of Genesis as requiring that Adam be the first human. Walton describes it this way:

Paul does not demand that Adam and Eve are the first or only humans. When he speaks of Adam as the "first man," he is most interested in the archetypal role of Adam and in the theological issues surrounding sin.

Even if one were to insist that Paul was truly identifying Adam as the first human being on earth, something that does not appear to be necessary, there remains the possibility that he was simply wrong on this point. As described in Chapter 1, the Bible is demonstrably wrong in a number of cases, and Biblical characters sometimes believed things that were not true. For instance, in the 6th Century BCE the prophet Daniel declared that the king of Babylon Nebuchadnezzar controlled every kingdom on earth by God's decree.

> *Your Majesty, you are the king of kings. The God of heaven has given you dominion and power and might and glory; in your hands he has placed all mankind and the beasts of the field and the birds in the sky. Wherever they live, he has made you ruler over them all. (Daniel 2:37-38)*

This was certainly not true as he had no influence over the nearby Median Empire in what is now Turkey, Iran, and Afghanistan, not to mention contemporary empires further away in Asia, Europe, and Africa.

Paul himself made a statement regarding head coverings and hair length that he clearly felt strongly about, but which is generally dismissed by most modern Christian leaders.

> *4 Every man who prays or prophesies with his head covered dishonors his head.*
>
> *5 But every woman who prays or prophesies with her head uncovered dishonors her head - it is the same as having her head shaved.*
>
> *8 For man did not come from woman, but woman from man;*
>
> *9 neither was man created for woman, but woman for man.*
>
> *13 Judge for yourselves: Is it proper for a woman to pray to God with her head uncovered?*
>
> *14 Does not the very nature of things teach you that if a man has long hair, it is a disgrace to him,*
>
> *15 but that if a woman has long hair, it is her glory? For long hair is given to her as a covering.*
>
> *16 If anyone wants to be contentious about this, we have no other practice - nor do the churches of God. (1 Corinthians 11)*

Notably, Paul appeals to the creation of Adam and Eve in his insistence that women must wear head coverings when praying and that men must not have long hair. Moreover, his language suggests that he did not believe that his views were specific to his time and culture since he asserts that they are grounded "in the very nature of things" and insists that they apply in every single church. In the same way that he was mistaken about the universal applicability of these views, it is conceivable that he could have been wrong in his views that Adam was the first created human being on the earth.

Finally, we must consider the knowledge base of our ancient forebears and ask: What else should we expect them to have believed? There was no such thing as scientific inquiry at that time, and there would have been no reason for God to reveal the true nature of creation to them as it was not necessary for their spiritual health. Just as God did not reveal the details of deep time and biological evolution to them, neither do we see any evidence in the Bible or in any other ancient documents that God revealed to anyone the true nature of stars and planets or how an egg and sperm combine to form a new human being. Instead, He let them believe whatever they wanted to believe and worked to focus their attention on those matters of faith that truly mattered.

The arrival of humanity in Eden

The two phases of creation that are necessary under a Revised Origins Paradigm - the creation of humanity generally and the appearance of Adam and Eve specifically - are not distinguished in the Bible in an obvious way. Some have taken the two different creation narratives in the first two chapters of Genesis as implicitly making this distinction:

> *Creation of all humanity*
> *So God created mankind in his own image, in the image of God he created them; male and female he created them. (Genesis 1:27)*

> <u>*Creation of Adam*</u>
> *Then the LORD God formed a man from the dust of the ground and breathed into his nostrils the breath of life, and the man became a living being (Genesis 2:7)*

The difficulty is that the description in Genesis 2:7 of a process whereby raw materials are reorganized into a human body is used as evidence of Biblical support for biological evolution, something that should only be relevant in Genesis 1:27 if that is where the creation of all humanity is described. This muddles the attempt to see Chapters 1 and 2 as supporting a two-phase creation.

Still, the existence of two different phases of creation does not require that the Bible spell out those phases. Christ's entrance into the world is also in two phases, the suffering servant and the yet-to-be-realized conquering king, but before Jesus walked the earth there was essentially no theology of a two-phase process. One might also point to the nature of the Trinity as a theology that is not precisely characterized in the Bible and took centuries to fully develop, and has since been treated as orthodox.

For the purposes of this book, we can fast forward through the first phase of creation. As of 2017, the oldest known fossil belonging to Homo sapiens is 300,000 years old and was found in Morocco (per Strickland). Modern human fossils aged at least 177,000 years have been found in Israel, and evidence of human occupation to the east in Mesopotamia has been found that dates back 80,000 years. In the ancient region of Sumer, in what is now southern Iraq and Kuwait, the earliest human artifacts date to around 6,500 BCE. Based on patterns of human migration from Africa through what is now Saudi Arabia, Syria, Turkey, and Iran, it is likely that humans lived in Sumer considerably earlier than 6,500 BCE. However, much of the history of this region is obscured underneath sediments as flooding was common.

There is a range of estimates under the Traditional Origins Paradigm for the age of the earth and the creation of Adam and Eve of between 5,000 and 8,000 years, based on different methods for filling in the gaps in the Biblical timeline. The most commonly cited among them is about 6,000 years, placing the creation account at about 4,000

BCE. In Mesopotamia where Eden is believed to have resided this timeframe is called the Ubaid period. Archaeologically, it is characterized by a transition from smaller villages to larger cities with associated social stratification, advances in metallurgy and irrigation, and the development of trade. Mesopotamia in the Ubaid period is where civilization was born, and it was there that the story of Adam and Eve began.

Eden was a large region in what is now the middle east, and the garden where much of the events of Genesis Chapters 2 and 3 took place was a smaller area within Eden. The Bible describes the location of Eden by identifying four rivers. Two of those rivers - the Tigris and Euphrates - retain their names today and thus provide solid landmarks, while the other two - Pishon and Gihon - cannot be identified with any certainty. Complicating the hunt for the precise location of Eden and of the garden within it is the fact that the geography of the region has changed considerably over time. For instance, at the time of Adam and Eve's appearance, the Persian Gulf extended an additional 200 km inland as compared to today. The fact that the region exhibits strong evidence of repeated flooding also means that the Pishon and Gihon rivers may have shifted course over the last six millennia or may even no longer exist. While there are multiple theories about the area that the Bible is identifying, the predominant view seems to be that Eden covers the region between and around the Tigris and Euphrates rivers, and that the garden within Eden most likely lay in Sumer at the southeast end of Eden, near the Persian Gulf.

Archaeological evidence gives us a rich story of human habitation and culture in Eden prior to and around the time of Adam and Eve's appearance. As far back as 10,000 BCE there is evidence of agriculture in the region, and over the following millennia it matured to include plowing and irrigation. By 6,000 BCE, the earliest time estimated for the appearance of Adam and Eve, animal husbandry was common, and long-distance trade had begun.

In about 5,400 BCE, Eridu was founded in Sumer, the first large settlement in the region. Eridu would grow to become a significant urban area and is credited by many with being the birthplace of true civilization. Indeed the ancient Sumerians believed that Eridu was the first city in the world, and they may have been correct. By 4,000 BCE, the time at which most traditionalists estimate the creation of Adam and Eve to have occurred, multiple other towns had arisen in Sumer, including Bad Tibira, Uruk, Ur, and Girsu.

Eridu's primary economic driver was fishing as evidenced by the remains of fishing nets and gear. Models of reed boats have also been found at the site and are the earliest indication of the construction of boats anywhere on earth. But Eridu had other industries as well, including subsistence agriculture irrigated by constructed canals, and large scale herds of sheep and goats. Eridu was also a major pottery production center, as multiple kilns and significant amounts of pottery fragments have been unearthed.

The "creation" of Adam and Eve

Eridu is a reasonable candidate for the birthplace of Adam and Eve due not only to its location in Sumer and the fact that it was one of largest cities in the region, but because Eridu was the locus of creation in Sumerian mythology and its patron god Enki was the creator of humankind. Enki would also play a prominent role in the Sumerian flood story with remarkable parallels to the story of Noah and the ark in Genesis Chapters 6 - 8. Of course, Enki had little in common with the God of the Bible, being one of many gods within the Mesopotamian pantheon and being known for trickery and sexual exploits. But Adam and Eve's exposure to and schooling in a god of creation may have prepared them for an encounter with the true Creator God.

There is no reason to think that Adam and Eve were in any way special, but they did have a special calling. God called Adam and Eve out of their ordinary lives into something - and somewhere - extraordinary. Figuratively, God "created" them in the broad region called Eden and then subsequently placed them within the garden.

> *Now the LORD God had planted a garden in the east, in Eden; and there he put the man he had formed. (Genesis 2:8)*

Interestingly, the Sumerian word for the region was Edin despite Sumerian being distinctly different than the Semitic language which would ultimately evolve into Hebrew.

Though we can only speculate on the particulars of how Adam and Eve heard the call from God and journeyed from Eridu to the garden, Abraham's calling and journey provides a convenient model with numerous parallels. Like Adam and Eve, Abraham was a polytheist (according to Joshua 24:2) who lived in Ur for some period of time before his father Terah took he and his brothers northwest to Haran. Ur was a mere 30 miles away from Eridu and operated as its twin, with Ur being the political center to Eridu's religious center.

The Bible gives us no insight into how Abraham (then Abram) heard God's call, whether by a vision, an audible voice, a visitation, or an unction. We also have no way to know what Abraham understood of the God who was calling him. Did he think it was one of the Mesopotamian pantheon, or did he understand that this was the Lord over all? There is some indication that Abraham's initial understanding of God was limited. In the story of Abraham's encounter with Melchizedek, following Abraham's defeat of the kings who had kidnapped his nephew Lot, we see Abraham giving a tenth of all he owned to Melchizedek. This is the first record of a tithe to a priest, and it immediately followed Melchizedek's proclamation that Abraham served "God Most High, Creator of heaven and earth" (Genesis 14:19). Just a few verses later, Abraham uses nearly identical language in referring to "the LORD, God Most High, Creator of heaven and earth" (Genesis 14:22) and it is the first time that Abraham describes God in this way. It may be that Melchizedek revealed to Abraham the true nature of the God that Abraham served, and that before his encounter with Melchizedek Abraham did not truly understand that the God who had called him was really the Creator of everything and everyone and the only true God.

Adam and Eve's call by God may have occurred in a similar way. Whatever the particulars, we can imagine that they felt compelled to leave their family and homeland just as Abraham was called to do, and to set off to someplace new. And like Abraham, they may not have understood exactly who this God was who was calling them. For instance, after their disobedience in the garden, they attempt to hide from God (Genesis 3:8). This suggests that Adam and Eve may not have understood that God was omniscient. Nevertheless, Adam and Eve believed that the God who was calling them was a God worth following.

That Adam was born of a woman rather than being formed from dust may appear at first to be a dramatic departure from Scripture, but it need not be. Other passages refer to all humans being made from dust despite the fact that all are born from a woman:

> *for he knows how we are formed, he remembers that we are dust. (Psalm 103:14)*

> *All go to the same place; all come from dust, and to dust all return. (Ecclesiastes 3:20)*

Moreover, the Bible uses the word "create" to describe the creation of all humans beings.

> *Let this be written for a future generation, that a people not yet created may praise the LORD (Psalm 102:18)*

> *But now, this is what the LORD says - he who created you, Jacob, he who formed you, Israel: "Do not fear, for I have redeemed you; I have summoned you by name; you are mine. (Isaiah 43:1)*

The word for "create" in these verses (*bara*) is the same word used to describe the creation of mankind in Genesis 1:27. Thus the natural biological process of procreation can be treated as coincident with God's act to create mankind.

The Biblical description of Eve's creation is, however, less straightforwardly captured under the Revised Origins Paradigm. While there is no obvious way to tie the Biblical description of Eve's having been formed from one of Adam's ribs (or more precisely, from a part of his side) to the natural process of birth, it is possible to pair the intention behind the Biblical description to what occurs in the natural. Just after the description of Eve's creation, Genesis provides an editorial description of marriage:

That is why a man leaves his father and mother and is united to his wife, and they become one flesh. (Genesis 2:24)

This verse suggests that there is a connection between the way that Eve was created and marriage; Adam and Eve were to come together and become "one flesh" in an apparent reversal of the fact Eve had just been separated from Adam's flesh. It is as if man and woman were intended to be both separate entities as well as a single entity, a condition we should recognize as mimicking the Trinity. This parallel becomes even stronger when we consider Adam's words upon seeing Eve for the first time.

The man said, "This is now bone of my bones and flesh of my flesh (Genesis 2:23a)

Despite being a separate individual, Eve was of the same substance as Adam. The Trinity can be described in the same way: while each of the three Persons are distinct, they are of the same essence, a condition called homoousion.

The fact that Eve was meant by God to be a counterpart to Adam to reflect the Trinity is also apparent in another way. The editorial statement that they would be "one flesh" is an indication not only of unity but also of sex, and thus procreation. Just as a man and woman come together to produce a child, so also the three persons of the Trinity worked together to produce humanity.

Let us make mankind in our image, in our likeness (Genesis 1:26)

The Revised Origins Paradigm incorporates these features of Eve's creation and God's purposes behind it, while departing from the literal description of Eve being formed from flesh taken from Adam.

The uniqueness of humanity

Adam and Eve may not have been unique among their contemporaries, but the Bible makes humanity as a whole unique among all earthly creatures. The Bible presents us with a special creation of Adam and Eve and simultaneously declares that they were created in the image and likeness of God. This presentation sets humans apart from all other creatures in an obvious way. But under a Revised Origins Paradigm wherein humans emerge from a biological evolutionary process spanning millions of years, it is not as straightforward to see how humans could be unique. There is a temptation to see humans as different from apes only in degree rather than in kind, more advanced but of the same stock.

Humans do in fact stand apart from our closest cousins the chimpanzee and bonobo in several crucial ways. For instance, we differ in our use of symbolic and abstract thought and the expression of such thoughts in language. It is this characteristic that makes us capable of believing in an unseeable God. Add to this our understanding of moral absolutes and the concept of right and wrong apart from immediate ramifications, and we have the ability to be obedient to God as well. There is no indication that other apes are capable of such things. They may use some form of language, but the components of their language are always concrete things, that which can be identified through the five senses or associated with an emotion. Apes can abide by a rudimentary code of conduct, but their behavior is driven by the anticipated outcomes of their actions (for instance, whether they can remain within their group with peaceable relationships) rather than by any abstract principle of the goodness or badness of those actions.

Beyond abstract and symbolic thought, language, and morality, we understand ourselves to differ from apes in having a spiritual nature which is conferred to us not through biology or biological evolution but rather directly from God, and which will endure after our physical bodies die and decompose. Our uniqueness in this regard is generally attributed less to natural observations and more to the Bible's declaration that we are made in the image and likeness of God, though most other religions and many philosophers over the millennia have presumed that humans have an immortal spiritual component.

The Bible differentiates between the soul and spirit of a human being, but is not always clear about the distinctive character of each.

That the soul and spirit are different is clear enough from verses such as these:

> *May God himself, the God of peace, sanctify you through and through. May your whole spirit, soul and body be kept blameless at the coming of our Lord Jesus Christ. (1 Thessalonians 5:23)*

> *For the word of God is alive and active. Sharper than any double-edged sword, it penetrates even to dividing soul and spirit, joints and marrow; it judges the thoughts and attitudes of the heart. (Hebrews 4:12)*

We can infer that the soul is the seat of the personality, the essential self. This is most apparent when comparing the treatment of the word for soul (*psuche*) in two different translations.

> *And I will say to my soul, Soul, you have ample goods laid up for many years; relax, eat, drink, be merry. (Luke 12:19, ESV)*

> *And I'll say to myself, "You have plenty of grain laid up for many years. Take life easy; eat, drink and be merry." (Luke 12:19, NIV)*

Consistent with this definition, it should be no surprise that even God has a soul.

> *Behold, my servant whom I have chosen, my beloved with whom my soul is well pleased. I will put my Spirit upon him, and he will proclaim justice to the Gentiles. (Matthew 12:18, ESV)*

As the seat of the personality, the soul should embody our mind, will, and emotions. However, the Bible is not clear on this point. In some cases we see emotions directly associated with the soul.

> *Then he said to them, "My soul is overwhelmed with sorrow to the point of death. Stay here and keep watch with me." (Matthew 26:38)*

In other cases, emotions and thoughts appear to be something separate from the soul:

> *Love the Lord your God with all your heart and with all your soul and with all your mind and with all your strength. (Mark 12:30)*

The human spirit, on the other hand, is less ambiguous. Most importantly, it is the immortal component of every human being.

> *and the dust returns to the ground it came from, and the spirit returns to God who gave it. (Ecclesiastes 12:7)*

> *While they were stoning him, Stephen prayed, "Lord Jesus, receive my spirit." (Acts 7:59)*

Being immortal, the spirit is the means through which we commune with the eternal God.

> *The Spirit himself testifies with our spirit that we are God's children. (Romans 8:16)*

The spirit is also the source of our desire to be faithful to God.

> *Watch and pray so that you will not fall into temptation. The spirit is willing, but the flesh is weak. (Matthew 26:41)*

But the Bible also presents some conflicting associations between spirit and soul. For instance, in some places the Bible suggests that the spirit and mind are separate while in others it suggests that they are the same.

> *For if I pray in a tongue, my spirit prays, but my mind is unfruitful. (1 Corinthians 14:14)*

> *and to be renewed in the spirit of your minds. (Ephesians 4:23, ESV)*

The Bible also uses "spirit" in places where it would seem that "soul" would be more appropriate insofar as emotions are being addressed.

> *After he had said this, Jesus was troubled in spirit and testified, "Very truly I tell you, one of you is going to betray me." (John 13:21)*

> *my spirit was not at rest because I did not find my brother Titus there. So I took leave of them and went on to Macedonia. (2 Corinthians 2:13, ESV)*

It would appear, then, that the human soul and human spirit are not altogether separate entities, but are connected in some way. This makes sense when we consider that it is the spirit, not the soul, which is immortal, and yet the soul also endures in the afterlife (it must, since otherwise our personalities and unique selfhood would be erased upon our death). It is as if the spirit preserves the soul; without the spirit, the soul would fade into nothingness when the body dies.

If modern humans are the result of biological evolution rather than an instantaneous creation, we must confront the question of how we came to be in the image and likeness of God, having a spirit in addition to a soul. We might separate the two in this effort, relegating the soul to the purely biological and presuming that, at some point in our history, God added a spirit to our ancestors. Collins and Cole-Turner discuss this possibility, and it would be consistent with the fact that the higher animals also have minds and emotions and thus can be said to have a soul. But this approach appears to be at odds with the conclusion that the soul and spirit and connected, and with the fact that the human soul has distinctive characteristics that set it apart from that of apes.

The compromise, it would seem, is that God imbued our ancestors with spirits prior to the final stage in the evolutionary development of our souls, or alternatively to inaugurate that final stage. Those human spirits, connected as they were with the Spirit of God, channeled the development of souls to add the elements that set us apart from all other animals: abstract and symbolic thought, language, and morality. We can imagine that the first hominids to have an immortal spirit grew to recognize the presence of that spirit and to believe in God only over long periods of time, somewhere between generations and epochs. It would be a process not unlike the development of a modern human baby into an adult. The baby has a spirit but has no clear sense of self, no ability to form abstract thoughts, no language skills, and no conscious recognition that God exists. In order for its spirit to have any true relevance for that baby's experience of earthly reality, its soul and the biological systems that support it must grow over time. Similarly, the souls of our hominid ancestors would have had to develop to the point where their spirits could become relevant for their experience of reality, and to make them consciously capable of relating to God.

It would be ideal if there were within the record of human evolution a clear demarcation between non-human and human, or between humans of different moral and spiritual capabilities. No such boundary has been discovered, which may be unsurprising given that those capabilities may not have left any evidence in the ancient bones and artifacts that are available to us. Even so, we know that such a transition occurred even if we cannot say precisely when or how quickly, insofar as there is strong evidence of a divergence in ancestral hominids about 6-7 million years ago, with one of the paths leading to modern humans with immortal spirits and distinctive souls and another leading to modern apes.

We can, however, make some educated guesses. Archaeologists have discovered burial sites dating back about 100,000 years whose contents suggest the possibility of belief in an afterlife. This would suggest that the first humans with spirits appeared at least 100,000 years ago. Evidence regarding brain size of hominids and tool-making suggest that the transition to souls having something like modern capabilities could have occurred between 1.5 and 2.5 million years ago. This also corresponds to the period when genus Australopithecus gave way to the genus Homo to which we modern humans belong. Admittedly this range is large and uncertain, and may or may not become more precise with further research.

How would God have related to the earliest hominids to whom He had given a spirit? Without a mind capable of imagining an unseeable God, those hominids would have had only the most nebulous of notions of the spiritual realm, but this would not have stopped God from treating them as His beloved children. It would also not necessarily have stopped them from being taken to heaven when they died. The circumstances may have been little different than a modern human baby that dies prematurely. Theologians have long believed that the baby's inability to understand spiritual matters and to choose good over evil is not a barrier to its entrance into heaven, based on passages such as this:

> *Why did I not perish at birth, and die as I came from the womb? Why were there knees to receive me and breasts that I might be nursed? For now I would be lying down in peace; I would be asleep and at rest (Job 3:11-13)*

The same would be true for any human being born with severe mental retardation. God is all the more gracious and merciful to those least able to turn to Him.

Humanity's relationship with God prior to Adam and Eve

Regardless of their views on the origins of humanity, most Christians generally view the Bible as providing the full story of God's relationship with humanity. Every point in history at which God interacted with humanity in any consequential way is recorded, and there is nothing missing that would be of relevance to God's redemptive work. While this may be true, the Bible also provides several hints that God interacted with humanity in meaningful ways that are not recorded in any detail. These hints give us a basis for asserting that God interacted with humanity before Adam and Eve.

After Abraham's defeat of the kings who had kidnapped his nephew Lot, he is met by the king of Sodom and the priest Melchizedek.

> *Then Melchizedek king of Salem brought out bread and wine. He was priest of God Most High (Genesis 14:18)*

From this verse we can deduce that there was already a priesthood in Salem (which would later become Jerusalem) centuries before God established the priesthood through Aaron. Clearly, God had given some form of revelation of Himself to people not associated with

Abraham, and the people had developed some system of service to Him. We know nothing of the nature of this priesthood, but we can be sure that there is a story behind its inception and development that is not covered in the Bible.

Later, after the Israelites enter the promised land, we see what appears to be a prophet of God that was similarly not associated in any way with Israel. Knowing of the previous victories that the Israelites had had over their enemies, Balak the king of Moab sends an envoy to Balaam asking him to curse the Israelites for him. Balaam responds by saying:

> *"Spend the night here," Balaam said to them, "and I will report back to you with the answer the LORD gives me." (Numbers 8a)*

We then see an exchange between Balaam and God that has the character of dialogue that we see later in the Bible with Hebrew prophets: direct questions and unambiguous answers. Balaam even refers to God as "the LORD my God." Upon meeting Balak, Balaam describes his submission to God's direction:

> *"Well, I have come to you now," Balaam replied. "But I can't say whatever I please. I must speak only what God puts in my mouth." (Numbers 22:38)*

Once again, this is an indication that God was at work in people outside of the primary storyline of the Bible. There is no reason that this could not also have been true for humanity before the time of Adam and Eve.

As humanity evolved to recognize the spirit that God had placed within them, they sought to connect with the spiritual realm. Their efforts led them to imagine deities that were responsible for different aspects of the world in which they lived. But without direct revelation from God, they did not know that there was only one true God, and polytheism dominated the religious landscape. In Eridu, the local god Enki was worshipped in a series of seventeen temples, each built over the remains of a previous one, with the earliest dating back to about 4900 BCE. Enki was believed to live in the primeval aquifer far below the earth which provided all the fresh water on which the people of Eridu depended for fishing and agriculture. Elsewhere in Sumer, local gods were associated with other aspects of nature such as wind and storms (Enlil in the city of Nippur), the sky (An in the city of Uruk), and the moon (Nanna in the city of Ur).

When Paul and Barnabas were preaching in Lystra, residents concluded that they were members of the Greek pantheon based on the miracles they performed there. In response, Paul and Barnabas attempted to correct their thinking and direct the city's residents to the one true God who was responsible for creating everything, as opposed to Greek mythology where different gods were responsible for different elements of creation:

> *Friends, why are you doing this? We too are only human, like you. We are bringing you good news, telling you to turn from these worthless things to the living God, who made the heavens and the earth and the sea and everything in them. (Acts 14:15)*

Paul then makes a statement wherein he appears to give some amnesty to those who adhered to other religions in the past.

> *In the past, he let all nations go their own way. (Acts 14:16)*

Given Paul's audience, he is most certainly referring to polytheistic religions at minimum, and potentially to all ways of life that are askew of faith in and faithfulness to the one true God. Paul is hinting at a progressive unfolding of truth about God that, in times before his own, had only been available in some rudimentary form. In his letter to the Romans, we see Paul make a similar statement about those who were not privy to the full truth about the cross.

> *in his forbearance he had left the sins committed beforehand unpunished. (Romans 3:25)*

It would seem that, according to Paul, peoples to whom less had been revealed about God had less responsibility in how they related to God. In the same letter, Paul expounds on the idea that God expects us to respond in a manner consistent with the nature of the revelation given to us:

> *Indeed, when Gentiles, who do not have the law, do by nature things required by the law, they are a law for themselves, even though they do not have the law. They show that the requirements of the law are written on their hearts, their consciences also bearing witness, and their thoughts sometimes accusing them and at other times even defending them. (Romans 2:14-15)*

For humans who lived before the time of Adam and Eve, the same would have been true. Those individuals who endeavored to act righteously in accordance with the truth as they understood it, even if in the context of a polytheistic religion, would have been treated mercifully by God.

Was Adam perfect?

The Traditional Origins Paradigm characterizes pre-Fall Adam as being perfect in every way, physically, intellectually, morally, and ethically, and of course sinless, a paragon of what all humans were meant to be. This view can be traced back to Ambrose in the late 4th Century. But there is no indication that this was a common view prior to Ambrose. This is not surprising as there is nothing explicit in Genesis to support it.

One verse that might appear to support the traditional view is this:

> *This only have I found: God created mankind upright, but they have gone in search of many schemes. (Ecclesiastes 7:29)*

Under the traditional view "mankind" was composed of only Adam and Eve during the time of the events in the garden of Eden, and thus this verse is a reference to their perfect nature at the time of their creation. But this interpretation would make this verse unique in Ecclesiastes given that there is nothing else in the entire book which references Adam and Eve or God's initial creation of the world. Indeed, the broader context of the book is clearly humanity as a whole and the author's contemporaries most particularly. Moreover, as discussed above, the creation of which this verse speaks could easily refer to all human beings rather than only Adam and Eve. This verse is thus more likely to refer to God's intentions for every human being rather than their initial moral state; God created mankind with the intention that they lead upright lives, but they have instead pursued evil.

Another verse often used in support of the view that Adam was created perfect appears to identify his essence with the essence of God.

> *Then God said, "Let us make mankind in our image, in our likeness... (Genesis 1:26a)*

As God is perfect, so the argument goes, man made in His image must also have been perfect. But the second half of this verse actually provides a possible reason that God created mankind in His image:

> *... so that they may rule over the fish in the sea and the birds in the sky, over the livestock and all the wild animals, and over all the creatures that move along the ground. (Genesis 1:26b)*

It appears that the image and likeness of God is identified with governing authority, not Adam's moral nature. Adam was created to be a ruler just as God is a ruler. This possibility is bolstered by a statement made by Paul in which he associates man's vocation as a ruler with God's purpose in creating him.

> *But there is a place where someone has testified: "What is mankind that you are mindful of them, a son of man that you care for him? You made them a little lower than the angels; you crowned them with glory and honor and put everything under their feet. (Hebrews 2:6-8a)*

Even if one were to associate Adam's purpose in ruling over all animals with the fact of his creation rather than the fact of his having been made in the image and likeness of God, there would still be no need to associate the image and likeness of God with a perfect moral character. There are, in fact, a wide variety of possibilities as Collins points out.

> *There is no unanimity among Biblical scholars on just what the image and likeness of God means.*

Some of the possibilities are directly suited to Adam under a Revised Origins Paradigm. If image and likeness must refer to something of God's attributes, one possibility is that Adam was "crowned with glory and honor" as described in the Hebrews passage above, just as God is crowned with glory and honor. Another possibility is that image and likeness refers to God's power to do as he wills as argued by Tertullian in the early 3rd Century.

> *the prohibition [against eating from the tree of knowledge] relied on and proved the gift of freedom, which (together with personal initiative (potestas)) is the hallmark of the divine image and likeness. (per Bouteneff, brackets added)*

It is also possible that humans may be like God in being triune. Since God refers to Himself in the plural when saying that He created man "...in our image, in our likeness," He may be referring to the fact that we are in some parallel way triune in that we are composed of body, soul, and spirit. Some have taken this concept a step further and associated the Trinity speaking in Genesis 1:26 with the next statement in Genesis 1:27 that God created mankind "male and female." In this case, it is the union of man and wife in marriage (two parts to one whole) which is an image and likeness of God's triune nature, albeit it in an inferior way.

Others have made a different connection between mankind being created male and female and also in the image and likeness of God. As God is a community unto Himself in the context of the Trinity, so our relationship with one another creates a community where we are called to relate to one another with the same unity and love that characterizes the Trinity. In this context, the creation of humans as "male and female" is a reference not to sexual differentiation, but to this broader community. God's goal of creating a community is also reflected in His declaration in Genesis 2:18 that it was "not good" that the man was alone.

Finally, there is the view that humans beings created in the image and likeness of God refers to the fact that we have immortal spirits just as God is immortal. And as a consequence, humans are the only creatures that can have a meaningful and conscious relationship with Him, which was God's ultimate purpose in creating mankind.

Under the Traditional Origins Paradigm, the idea that Adam was created perfect is not only crucial to the doctrine of original sin, but has been held by all orthodox Christians since at least the time of Jesus. But the doctrine of original sin in its modern form did not exist prior to the 4th Century as will be discussed in more detail in Part III, and Adam's supposed perfection was not universally held even at that time. For instance, Cyril of Alexandria believed that the created cannot be expected to fully exhibit the characteristics of their Creator.

> *All things were brought into being by God, and he himself is endowed with a nature of a different kind from that of creatures. For that which has been made is subordinate to its Maker, and by its nature is in every way inferior to it. (per Russell)*

Clement of Alexandria addressed Adam's supposed perfection more explicitly, saying that he

> *was not created perfect in constitution, but suitable for acquiring virtue (per Kelly)*

When Paul addresses the practice of head coverings in the church, he indirectly addresses the connection between being made in the image of God and being perfect.

> *A man ought not to cover his head, since he is the image and glory of God (1 Corinthians 11:7a)*

If post-Fall man, imperfect as he is, can still be described as being the image of God, we must admit the possibility that being created in the image of God does not require Adam to have been perfect.

Conclusion

While the Traditional Origins Paradigm holds that Adam and Eve were the first human beings to have existed and the only humans on earth at the time of the events in the garden of Eden, the Bible is not explicit on these points. It describes the creation of two specific human beings, but does not proclaim that they were the first human beings. The absence of any mention of other humans cannot be taken as evidence that they did not exist. Those others may simply have not been relevant to the story of God's first meaningful interaction with humanity.

As in other parts of the Bible, the creation of Adam and Eve contains both literal and figurative elements. Their creation was literal in the sense that Adam and Eve were real people called by God to play an essential role in His plan to reach out to mankind. It was figurative in the sense that Adam was not formed directly from the dust of the ground and Eve was not formed from one of Adam's ribs. Instead, they both arose from ordinary procreation. The Biblical description of their appearance on earth was meant to convey God's role as Creator and Father as well as the nature of the relationship between Adam and Eve. This blending of the literal and the figurative is similar to what we see in John's vision of heaven.

> *Then I saw a Lamb, looking as if it had been slain, standing at the center of the throne, encircled by the four living creatures and the elders. The Lamb had seven horns and seven eyes, which are the seven spirits of God sent out into all the earth. (Revelation 5:6)*

There is every reason to believe that the throne, the living creatures, and the elders were all real, but the image of Jesus as a lamb which was slain and which had seven horns and seven eyes is clearly meant to be figurative. In inspiring the words that became the book of Genesis, the Holy Spirit likewise determined that a combination of literal and figurative elements was the best way to describe the beginning of the story of His work to bring mankind to Himself.

Chapter 7
Adam and Eve in the Garden of Eden

The Biblical story of Adam and Eve in the garden of Eden is just as true under the Revised Origins Paradigm as it is under the Traditional Origins Paradigm. The garden is a special place designed by God for Adam and Eve where they enjoy His favor and blessings. God assigns them a vocation in the garden and places boundaries on what they are and are not allowed to do. Adam and Eve cross those boundaries and are consequently punished by God. And that punishment has had profound impacts on all of humanity ever since.

But under the Revised Origins Paradigm, some of the details of the story of Adam and Eve in the garden must be taken in less literal terms than under the Traditional Origins Paradigm. In principle this is not a deviation from the traditional approach, as even there some details are taken figuratively as discussed in Chapter 1. In this respect, the difference between the two paradigms lies not in whether Genesis Chapters 1-3 in whole should be taken literally or figuratively, but rather what parts should be taken literally and which parts figuratively.

Entering the garden

According to the Bible, Adam was not created within the garden of Eden, but rather outside of it. God then moved him in to the garden.

> *Then the LORD God formed a man from the dust of the ground and breathed into his nostrils the breath of life, and the man became a living being. Now the LORD God had planted a garden in the east, in Eden; and there he put the man he had formed. (Genesis 2:7-8)*

There is no explanation for why God would do this, nor is there any indication as to the distance between the location of Adam's creation and the garden. Regardless, it carries no weight under the Traditional Origins Paradigm. Under the Revised Origins Paradigm, however, it is important. Adam was born of a woman in a community of people, and then at some point when he was capable of it, he travelled to the garden under the guidance of the Holy Spirit. Presumably the garden was some distance away from his hometown as it was both lush and devoid of other humans.

There is no reason to think that God transported Adam to the garden Himself. Certainly God is capable of such a thing, as with Philip's translocation in Acts 8:39-40. But the phrase "there he put the man" in Genesis 2:8 could simply mean that God arranged the circumstances so that Adam would get to the garden. We see similar phrasing in other places:

> *Observe the month of Aviv and celebrate the Passover of the LORD your God, because in the month of Aviv he brought you out of Egypt by night. (Deuteronomy 16:1)*

> *the LORD delivered me into your hands, but you did not kill me. (1 Samuel 24:18b)*

In neither case do we read the divine activity to be one of God moving people Himself. Rather, God works in His unfathomable way to ensure that those people arrive where He wants them to be even though they move under their own power and according to their own wills.

God's purposes in putting Adam and Eve in the garden

It is clear that the garden was a unique place within the region of Eden, and not only because God choose to put Adam and Eve there. As described in Chapter 2, the garden was particularly well suited for humans, being well-watered and having many edible plants. The world outside of the garden, in contrast, was less hospitable, a place where "thorns and thistles" thrived and where the ground was more difficult to work. The garden was a place where it would be easier to focus on God because the cares of daily life were less pressing.

Under the Traditional Origins Paradigm, Adam and Eve were in the garden for two reasons. First, God put Adam in the garden "to work it and take care of it" (Genesis 2:15b). Adam's first job was, apparently, as a gardener. This may have been a matter of subsistence if cultivation made it easier to produce sufficient edible plants, as opposed to merely foraging for them. If so, it would have been a natural enterprise for Adam, having been raised in an agrarian culture. But it could also have been a matter of establishing order in his new home. As described in Chapter 3, God may have completed His work on the sixth day, but He had not finished ordering His creation. For that, He enlisted humans. This would be consistent with God's broader purpose of assigning Adam and Eve the task of being His viceroys on earth.

Rule over the fish in the sea and the birds in the sky and over every living creature that moves on the ground. (Genesis 1:28)

The highest heavens belong to the LORD, but the earth he has given to mankind. (Psalm 115:16)

This purpose was not specific to the garden but rather to Adam and Eve's long term vocation, as described in more detail below.

God's second purpose for putting Adam and Eve in the garden, according to the Traditional Origins Paradigm, was to test them, to see if they would be obedient and continue to enjoy God's blessings, or would instead disobey God's command not to eat from the tree of knowledge and suffer the punishment of death. While this purpose could also apply under the Revised Origins Paradigm, it raises an uncomfortable question about the character of God: Why would He present Adam with a means to sin? Although there are other examples of God testing His children - for example, when He tells Abraham to sacrifice his son Jacob - in this case God's actions would seem to come perilously close to placing God in the role of tempter, something that we know cannot be true.

For God cannot be tempted by evil, nor does he tempt anyone (James 1:13b)

What good parent places candy in front of a child and then orders the child not to eat it?

The role of the tree of knowledge will be treated in more detail in Chapter 8 in the context of the discussion of the Fall. Here it is sufficient to say that it was no more a test than any of the Mosaic laws, which among other things were boundaries designed to keep the children of God safe. The tree existed not because God desired to set up a temptation, but because God intended to permit Adam and Eve to eat from it eventually. Knowledge is an integral part of earthly existence, and growth in knowledge is one of the destinies of all of God's children. As described by the anonymous author of the Letter to Diognetus in the 2nd Century,

> *For there cannot be life without knowledge any more than there can be sound knowledge without genuine life. So the two trees were planted close together. (per Louth)*

Just as God had knowledge of good and evil (Genesis 3:22), so also would His children eventually have such knowledge as they grew in maturity and in Godlikeness. But as attested to by several early theologians, Adam and Eve sought that knowledge before they had leave to do so. God's command not to eat from the tree of knowledge was not a test as much as it was a temporary restriction.

Under the Revised Origins Paradigm, there are two other reasons that God put Adam and Eve into the garden. The first was to refine and prepare them, to hone their Godlikeness. The second was an outgrowth of the first, a charge to be God's emissaries to all humans everywhere on the earth.

It is true that many early theologians believed Adam to have been created perfect, and this view persists today in the Traditional Origins Paradigm. But as described in Chapter 6, this view was not universal and there is a strong basis to believe otherwise. Gregory of Nyssa expressed his belief that being created in the image of God should not be taken to mean that Adam's initial state was one of maturity or completeness.

One who is made in the image of God has the task of becoming who he is. (per Louth)

In the 2nd Century, Irenaeus maintained that Adam was morally, spiritually, and intellectually a child. According to Kelly, Irenaeus's view could be summarized thus:

It was by a long process of response to grace and submission to God's will that Adam, equipped as he was with free choice, was intended to advance towards ever closer resemblance to his Maker.

Kelly also describes the similar views of Clement of Alexandria, saying that Adam

was childlike and innocent, destined to advance by stages towards perfection.

Genesis Chapter 1 describes humanity's purposes as inhabitants of the whole earth: they were to have dominion over all creatures and over the earth itself, and were to fill the earth and subdue it. God also tells them that every seed-bearing tree "on the face of the whole earth " is theirs for food. These purposes are clearly directed towards Adam and Eve's roles in the world writ large. There is no mention of the garden.

But then in Chapter 2 there is a turn. God does not send Adam and Eve out into the world. Instead, He plants a garden, puts Adam into it, and assigns him the task of taking care of it. Talk of the whole earth ceases, and descriptions of the location and specialness of the garden abound. One might think that the garden is where Adam and Eve are intended to spend the rest of their lives.

The transition from a whole earth perspective in Genesis 1 to a considerably smaller venue in Genesis 2 suggests that God was pursuing a stepwise process. While intending to eventually move them out to the whole world, He huddles first with Adam and Eve in a more intimate setting. It is not difficult to imagine, then, that their time in the garden was preparatory and temporary. They were not yet ready for the larger plan that God had for them. They needed to be trained in obedience, in taking care of God's creation, in fidelity in marriage, in trusting in God and in becoming more like Him. Then, when they were ready, they would finally be sent out of the garden, not as the exiled but rather as emissaries of God.

The tree of the knowledge of good and evil provides another reason to think that the garden was preparatory. As discussed above, the tree offered the same sort of knowledge that God had (Genesis 3:22), and so must have been something innately good. Rather than denying this knowledge to Adam and Eve for eternity, several early theologians believe that God was only denying it temporarily, until such a time as they were ready for it. If so, then it would appear that Adam and Eve's time in the garden was intended to help them grow in maturity and in Godlikeness, until at last they would have been given access to the tree of knowledge. A fuller treatment of the tree of knowledge is provided in Chapter 8.

The garden as a place of preparation gives rise to a new paradigm for God's purposes for Adam and Eve, one that hangs like an umbrella over the purposes described previously. This paradigm is Adam and Eve as the delegates of a new religion, one meant to supersede all others and provide the avenue through which all people would come to know the true God and enter into relationship with Him.

Adam and Eve as pseudo-Israel

The Bible records a long series of efforts by God to reach out to humanity and bring them into (or back into) a relationship with Himself. In nearly every case, His predilection has been to start with a single person or a small group, and use them as tools to begin something new in the world. We see it in dramatic fashion as Noah and his family are chosen to begin humanity anew when the rest of humanity is judged to be irredeemably wicked. We see it in the story of Abram to whom God reached out to become the patriarch of a great nation and to be a blessing to all peoples on earth. We see it with Moses and Aaron whom God uses both to free the Israelites from Egyptian bondage and to announce and implement God's law. We see it with Gideon, a farmer whom God raised up to be the leader of a rebellion against the Midianites and to reestablish his nation's fidelity to God. We see a similar pattern repeated with Samson, Samuel, and Nehemiah, and over and over again as God disciplines Israel but saves a remnant of the faithful to make a fresh start. And we see it in the life and ministry of Jesus, whose twelve disciples brought the gospel to Jews and Gentiles alike. One might also consider the many similar examples in the two thousand year history of Christendom, such as Saint Patrick, Francis of Assisi, and John Wesley. The great revivals of faith all began small.

Under the Revised Origins Paradigm, Adam and Eve were the first to receive this special calling from God to be examples of what it means to know and serve the one true God and to invite others to do the same. In doing so, they were to be the light of the world and the salt of the earth, just as Jesus proclaimed Israel to be in Matthew Chapter 5. In fact, there are multiple parallels between Adam and Eve and the nation of Israel. The parallels are so striking that it is difficult not to see Adam and Eve as a sort of pseudo-Israel, or even proto-Israel as Enns puts it. Collins and Walton make similar comparisons.

Israel's charge to influence the entire world for good is repeated multiple times throughout the Bible.

> *Observe them [decrees and laws] carefully, for this will show your wisdom and understanding to the nations, who will hear about all these decrees and say, "Surely this great nation is a wise and understanding people." (Deuteronomy 4:6, brackets added)*

> *In days to come Jacob will take root, Israel will bud and blossom and fill all the world with fruit. (Isaiah 27:6)*

> *I will make known my holy name among my people Israel. I will no longer let my holy name be profaned, and the nations will know that I the LORD am the Holy One in Israel. (Ezekiel 39:7)*

Israel was not to be insular and elitist, but rather to be outwardly focused. Unfortunately, examples of Israel actually acting in this capacity are few, as they repeatedly drifted from God and were in need of correction. This cycle of faithlessness, discipline, repentance, prosperity, and a return to faithlessness interrupted the larger plans that God had for them. But in some sense, they were repeating the failure of Adam and Eve, who likewise were intended to be a light to the nations and who likewise failed to fulfill that purpose. The similarity extends to the fact that Adam and Eve were exiled from the garden for their disobedience and just as Israel would be exiled to Babylonia in the 6th Century BCE.

Another parallel between Adam and Eve and Israel is in the similarity between the garden of Eden and the promised land. The garden was a geographic location given to Adam and Eve by God which had all the raw materials they would need, though they would still need to cultivate and care for it. Similarly, God gave the promised land to Israel, describing it as a land flowing with milk and honey (Exodus 3:8). This is also a reference to plentiful raw materials that nevertheless required some effort to gather.

Ephrem the Syrian went further, comparing the garden of Eden to Israel's tabernacle and the temple. The garden was a unique place of meeting between God and man, a link suggested by Ezekiel.

You were in Eden, the garden of God (Ezekiel 28:13a)

Likewise, the tabernacle and the temple would be the meeting place between the people of Israel and God. The most holy part of the garden would presumably have been where the tree of life resided, and possibly the tree of the knowledge of good and evil as well. This arrangement is reflected in the holy of holies in the temple where the ark of the covenant resided.

Dietary restrictions

With both Adam and Eve and Israel, God first established a special relationship with humans and then followed that action by establishing laws governing their behavior that would ensure the vitality of that relationship. In the case of Adam and Eve, tradition holds that there was only a single law: do not eat from the tree of knowledge. However, it is possible that there was another law with which Adam and Eve were expected to comply that has a conceptual counterpart among the laws given to Israel.

Before the commandment not to eat from the tree of life, God tells Adam and Eve that their food will be limited to plant material.

Then God said, "I give you every seed-bearing plant on the face of the whole earth and every tree that has fruit with seed in it. They will be yours for food." (Genesis 1:29)

The Traditional Origins Paradigm sees this verse less as a commandment than a statement of fact: Adam and Eve were vegetarians. But the Traditional Origins Paradigm also contends that Adam and Eve were created with all the knowledge and abilities needed for life in the garden, such as how to speak, how to tend a garden, and the meaning of the word "die." To be consistent, Adam and Eve would likewise have already been aware of the fact that they were vegetarians. But in that case God's statement would have been superfluous, the equivalent of telling them that water was to drink and air was to breath.

It seems more likely that, by specifying that plants were to be their food, God was implicitly prohibiting the eating of meat. Under the Revised Origins Paradigm, eating meat would have been within Adam and Eve's experiential background, and God's statement would therefore have had the quality of a commandment. In fact, Abel's offering hints at knowledge of eating meat that either came from his parents or from surrounding cultures.

In the course of time Cain brought some of the fruits of the soil as an offering to the LORD. And Abel also brought an offering—fat portions from some of the firstborn of his flock. (Genesis 4:3-4a)

Under the later law of Moses, the "fat portions" were considered the most desirable part of an animal to eat, and therefore represented the element of the offering having the greatest value. But if as required under the Traditional Origins Paradigm Adam and Eve had never eaten meat, had never even considered it, and there was no one else on earth who could be a meat eater, there would have been no reason for Abel to consider the fat portions as anything special. A better explanation for Abel's offering is that Adam and Eve were born into cultures where eating meat was commonplace and lived in a world where others ate meat, even if they themselves did not. They would have passed along to their children the fact that the fat portions of an animal sacrifice were highly valuable.

The purpose of God's commandment not to eat meat is not explicitly described in the Bible. But we do know that God later retracts this commandment, and this retraction provides some insight into its original purpose.

> *The fear and dread of you will fall on all the beasts of the earth, and on all the birds in the sky, on every creature that moves along the ground, and on all the fish in the sea; they are given into your hands. Everything that lives and moves about will be food for you. Just as I gave you the green plants, I now give you everything. (Genesis 9:2-3)*

In association with giving Noah and his family permission to eat meat, God paints a picture of humanity's relationship with animals that is decidedly different than that described during the creation narrative. Whereas Adam and Eve are told that they were to "rule" over all animals in Genesis 1:28, Noah is told that animals will "dread" humanity. This difference suggests that God's original intention was that humans have a respectful, custodial relationship with animals, something that changed to become more predator-prey in the time of Noah.

Even so, mankind was to continue to respect the animals used for food.

> *Do not cook a young goat in its mother's milk. (Deuteronomy 14:21b)*

This commandment is repeated in two other locations, and is part of a larger set of Mosaic dietary laws. It was intended to instill within Israel a sense that the creatures they used for food should nevertheless be respected as fellow created beings whose lives are not wholly different from humans. It would, in essence, be an afront to God to prepare a young goat for consumption by boiling it in the very milk that had given it life. The commandment given to Adam and Eve to be vegetarians and the commandment given to Moses are similar in that they both emphasize the implicit value of all creatures.

The full Deuteronomy verse from which the above text was quoted identifies another reason for the Mosaic dietary laws that could also be applicable to Adam and Eve under the Revised Origins Paradigm.

> *Do not eat anything you find already dead. You may give it to the foreigner residing in any of your towns, and they may eat it, or you may sell it to any other foreigner. But you are a people holy to the LORD your God. Do not cook a young goat in its mother's milk. (Deuteronomy 14:21)*

Israel was not allowed to eat any animal that died naturally, but could give it to a Gentile to eat. Presumably the animal in question was not putrid, or else this license would be tantamount to knowingly making a Gentile sick. More likely the animal in question was safe to eat, and there was nothing else in the Mosaic laws that would make the animal unclean and thus forbidden to be consumed. The fact that it was nevertheless prohibited as food means that the law was intended to make Israel stand out among the nations, to make it obvious that they were different. This fact is made apparent by the addition of "But you are a people holy to the LORD your God." The same reason is given at the end of Leviticus Chapter 11 after most of the dietary laws are given.

> *I am the LORD, who brought you up out of Egypt to be your God; therefore be holy, because I am holy. These are the regulations concerning animals, birds, every living thing that moves about in the water and every creature that moves along the ground. You must distinguish between the unclean and the clean, between living creatures that may be eaten and those that may not be eaten. (Leviticus 11:45-47).*

While there may have been some health benefits to the dietary laws, by and large there is agreement among theologians that their primary purpose was to distinguish between Israel and the surrounding Gentile nations. It may be that there was a similar purpose in God's commandment to Adam and Eve that they not eat meat. Vegetarianism would have been exceedingly rare in their time, so that they would have stood out among their contemporaries. Moreover, there is no reason to think that vegetarianism per se was God's intent, the above discussion on respect for animals notwithstanding, since Israel was permitted to eat meat. Jesus Himself could easily have reinstituted vegetarianism if it had been God's priority. Instead, vegetarianism for Adam and Eve appears to have been a means to an end, a way of marking them as God's own just as the Mosaic dietary laws did for Israel.

Conclusion

While Adam and Eve were placed by God in a flourishing garden within the region of Eden, it had never been His intention that they remain there forever, comfortable and insulated. Instead, they were there as preparation for their ultimate calling to be a light to the rest of the world. Just as He would later do with Abraham and others, God began a new enterprise with Adam and Eve wherein they would have a close and distinctive relationship with their heavenly Father and would grow in their knowledge of Him and their experience of serving Him. At some point when they were ready for it, God would have sent them out to engage other humans and to tell them of the joys and blessings of a life lived with and for God, perhaps retaining the garden as the home base to which they would regularly return.

Chapter 8
The Fall

The garden of Eden was intended to be home base for Adam and Eve's eventual forays back into the outer world from which they came, bringing the truth about the one true Creator God to the polytheists that surrounded them and calling them into a new relationship with Him. We might imagine their work in tending the garden actually expanding its borders so that it could support more and more believers. The garden would spread along with the number of occupants until the world was filled with not only its practical abundance but also the knowledge and presence of the Lord.

But this plan was interrupted almost as soon as it began. Adam and Eve failed to trust God completely in the face of half-truths offered by a deceiver, and they broke the command not to eat of the tree of the knowledge of good and evil. Their failure was the first in what would become a lamentable pattern of God reaching out to humanity over and over again to draw them into a relationship with Himself. The results of His efforts were mixed and the successes almost always temporary.

The central theme of the Fall - Adam and Eve's disobedience in the face of an unambiguous command from God - continues to apply under the Revised Origins Paradigm. The primary differences from the traditional view are in God's purpose for His command and in Adam and Eve's motivations in disobeying it.

The tree of knowledge

God's command to Adam that he was not to eat from the tree of the knowledge of good and evil (Genesis 2:17) implies that such knowledge was an unhealthy thing for Adam to have. But after Adam and Eve do eat from it, God says that they had become like Him in knowing good and evil (Genesis 3:22). The most straightforward reading is that Adam and Eve's newly acquired knowledge was exactly the same as God's knowledge. Nevertheless, the traditional view interprets the phrase "The man has now become like one of us, knowing good and evil" to mean that Adam and Eve's newfound knowledge was evil, and was the same as God's knowledge only superficially: Adam and Eve knew evil by experience whereas God knows evil only by observation. But the Bible makes no such distinction between the knowledge held by Adam and Eve and the knowledge held by God.

If God who is perfectly good has such knowledge, then we must conclude that that knowledge is intrinsically good. In fact in every other place in the Bible, knowledge is characterized as something worthy of acquiring, and it is often presented as being closely associated with wisdom.

> *For the LORD gives wisdom; from his mouth come knowledge and understanding; (Proverbs 2:6)*

> *The Spirit of the LORD will rest on him - the Spirit of wisdom and of understanding, the Spirit of counsel and of might, the Spirit of the knowledge and fear of the LORD (Isaiah 11:2)*

> *We continually ask God to fill you with the knowledge of his will through all the wisdom and understanding that the Spirit gives (Colossians 1:9b)*

Knowledge concerns facts and information whereas wisdom concerns understanding and discernment. One can have knowledge without wisdom, but wisdom requires knowledge. Despite its name, the tree of the knowledge of good and evil does not appear to have conferred knowledge only, but also wisdom. Eve correctly understood this when she recognizes that its fruit was

> *desirable for gaining wisdom (Genesis 3:6b)*

Thus the knowledge of good and evil that the tree offered represented the ability to distinguish properly between the two and to act accordingly. Having come from a polytheistic culture where ideas of good and evil would certainly have differed from those held by the true God, the ability to properly discern between the two would seem to be exactly the sort of skill that God should want Adam and Eve to have.

As discussed in Chapter 7, God's purpose in putting Adam and Eve into the garden was not to test them, but rather to prepare them. Preparation means that over a period of time one grows in the knowledge and skills needed to effectively complete some future task. For Adam and Eve, this task was to have been being God's emissaries to a world that did not know Him, and to accomplish this task they needed to understand His ways and be molded into His character. Becoming like God would have included receiving a divine revelation of good and evil of the sort that the tree of knowledge offered, so we can surmise that God would have allowed Adam and Eve to eat its fruit eventually, when He deemed them ready for it. Some early theologians expressed this very view. For instance, Gregory of Nazianzus said,

> *But it [the tree of knowledge] would have been good if partaken of at the proper time. (per Louth, brackets added)*

Irenaeus of Lyons said that without the knowledge of good and evil, one cannot even be properly human:

> *If you were to divest yourself of the knowledge of both [good and evil], and the double faculty of perception, you would divest yourself of the character of a human being. (per Bouteneff, brackets added)*

According to Bouteneff, Theophilus of Antioch took the position that knowledge in itself is not forbidden for human beings and there is nothing wrong with knowledge used properly and given at the proper time, but that Adam and Eve partook of it too early, in their infancy as it were.

The knowledge that the tree offered was good and to be desired, and appears to be something supernatural inasmuch as God has the same knowledge. This suggests that the knowledge in question was not a natural ability gained from the mere act of eating. Instead, it was a supernatural ability given by God where the fruit from the tree of knowledge was the means of transmission. This is not unique as there are other examples of God imbuing material things with special power. Most notably, the tree of life conferred immortality to anyone who ate from it (Genesis 3:22). But we also see it in the piece of wood that made bitter water sweet for Moses (Exodus 15:23-25), in the streaked, speckled, and spotted branches that produced streaked, speckled, and spotted goats and sheep for Jacob (Genesis 30:37-42), and in the physical strength given to Samson through his hair (Judges 16: 17-19).

Ironically, on this particular matter the Revised Origins Paradigm makes a more supernatural interpretation of the tree of knowledge than does the Traditional Origins Paradigm. Traditionalists hold that the tree held no special power or properties. It was a particular but ordinary tree that God chose only as a platform to test His children's fidelity to Him. The knowledge that Adam and Eve gained through eating its fruit was merely a realization of the wrong that they had committed; they came to a knowledge of evil by doing evil. Or, in a variant, Adam and Eve declared their right to decide for themselves what was right and what was wrong by rejecting God's version of right and wrong. They became like God not in being able to accurately discern between good and evil, but in discovering and exercising the independent power to make their own choices about what is good and what is evil.

The Traditional Origins Paradigm places significant weight on the sins of pride and moral self-autonomy as motivations behind Adam and Eve's disobedience. This perspective fits with the notion that the tree of knowledge was in no way special - its only purpose was to be forbidden. But as discussed further below, it is more likely that doubt and mistrust drove Adam and Eve's behavior. They disobeyed God, and that disobedience was a serious sin that deserved punishment, but they did not disobey because they were arrogantly rejecting God. Rather, they sought something special that the tree had to offer, something that any reasonable human being might legitimately have desired. That something was not just tasty fruit, but a new ability to see the world the way that God sees it. Only a tree containing special power from God would warrant that desire.

Obedience and maturity

God gave Adam and Eve no indication as to when, or indeed if, they would be given permission to eat from the tree of knowledge. Whatever its value, the tree was off limits. At the same time, God did not hide the tree of knowledge from them, nor prevent access to it. Arguably, He drew their attention to it. From this we can infer that God wanted Adam and Eve to be aware of the tree, but expected them to place a higher priority on obedience than on growing in knowledge.

> *The fear of the LORD is the beginning of knowledge (Proverbs 1:7a)*

Their situation was not unlike Joseph's dreams in which he saw his parents and brothers bow down to him (Genesis 37:5-10). Joseph did not know that the dream was about a time far off, long after he had been prepared for it. Similarly, Adam and Eve could see the tree of knowledge before them, but they did not know that its purpose was for a time far off, after they had been prepared for it. In his immaturity, Joseph told the dream to his family, and their angry reaction was predictable. He had taken matters into his own hands instead of keeping silent and trusting God to fulfill the dream in His own time. In the same way, Adam and Eve took matters into their own hands with regard to the tree of knowledge, and God's angry reaction was predictable. They had been expected to trust that God's prohibition against eating from the tree of knowledge was for their benefit, and that He would give them permission to eat from the tree if and when He deemed it appropriate.

Maturity must proceed commissioning, and obedience is a critical component of growing in maturity. Adam and Eve had a monumental task ahead of them, and they would not succeed if they were not properly prepared. Those who are impatient and move into positions of responsibility and authority before they have matured sufficiently are prone to pride that can disrupt their duties, damage those in their care, and tarnish the character of God. Paul's criteria for choosing bishops addresses this issue:

> *He must not be a recent convert, or he may become conceited and fall under the same judgment as the devil. (1 Timothy 3:6)*

Recent converts may know God, but rarely have the character needed to have spiritual authority over others or to be the embodiment of the love of God to those who are ambivalent or even hostile to Him. In the same way, Adam and Eve may have experienced great blessings from God in their calling into the garden and in the abundant provision they had there, but that alone was insufficient to make them capable emissaries.

Several early theologians expressed the belief that Adam and Eve were initially immature and would grow in maturity over time. For instance, Theophilus of Antioch referred to Adam as an infant, which Toews interprets to mean mental and spiritual immaturity rather than a literal newborn baby. Irenaeus of Lyons similarly suggested that Adam and Eve were more like children or adolescents than adults, immature forms of what was to come (per Cole-Turner), or persons in a state of unreadiness (per Bouteneff). Irenaeus also describes the process of growth and maturation that pre-Fall Adam and Eve were expected to follow:

> *Now it was necessary that man should in the first instance be created; and having been created, should receive growth; and having received growth, should be strengthened; and having been strengthened, should abound... (per Cole-Turner)*

Gregory of Nazianzus similarly saw Adam and Eve as a work in progress before the Fall.

> *So it [the tree of knowledge] was not beneficial for those still in a state of immaturity, greedy in appetite, just as mature food does not profit those who are still infants, still in need of milk. (per Bouteneff)*

Even if Adam and Eve had been created perfect as posited under the Traditional Origins Paradigm, there is reason to believe that they still would have needed some form of preparation before God sent them out into the world. Jesus Himself, while perfect, nevertheless needed a time of preparation before His ministry began. We see this preparation in His childhood where He learned to be obedient to His parents (Luke 2:51), we infer it for His early adult years prior to the beginning of His ministry when He lived in relative obscurity, and we see it in the temptation he underwent in the desert after being baptized by John (Luke 4:1-13). All the more should we expect that Adam and Eve, who under the Revised Origins Paradigm were not created perfect, needed a time of preparation.

> *The student is not above the teacher, nor a servant above his master. (Matthew 10:24)*

Their preparation included the difficult task of being obedient to a command that might have seemed unfair or even draconian after hearing the serpent's words.

It is worth pointing out the conundrum under the Traditional Origins Paradigm that Adam, though created perfect, nevertheless sinned. There is a clear conflict between Adam's complete inclination towards the good and his ultimate turn towards evil that has not been resolved to anyone's satisfaction, including traditionalists. Typically there is reference to the fact that Adam had free will and self-originated an evil inclination, but this begs the question of how such an evil inclination could have formed in one who was wholly pure. There is no such conundrum under the Revised Origins Paradigm, as Adam was an imperfect human being just as all his contemporaries were, and just as we are.

Satan and the serpent

There is no disagreement between the Traditional and Revised Origins Paradigms that it was Satan who deceived Eve. Although there is nothing in the Genesis account to connect the serpent with Satan, the book of Revelation makes this connection explicit.

> *The great dragon was hurled down - that ancient serpent called the devil, or Satan, who leads the whole world astray. (Revelation 12:9a)*

But the mode through which Satan worked in deceiving Eve is unclear. Did Satan literally cause an animal to speak, or did he communicate with Eve in some other way?

As discussed in Chapter 1, it appears certain that the serpent who at least appeared to be speaking was an animal rather than a bodiless voice or phantasm.

> *Now the serpent was more crafty than any of the wild animals the LORD God had made. (Genesis 3:1a)*

> *So the LORD God said to the serpent, "Because you have done this, cursed are you above all livestock and all wild animals! (Genesis 3:14a)*

If Satan spoke through the serpent, then, Satan must have possessed the serpent, taking over control of its soul and body. There is only one other unambiguous example in the Bible of demons possessing animals, and in that case the end result was the animal's immediate and devastating self-destruction.

> *When the demons came out of the man, they went into the pigs, and the herd rushed down the steep bank into the lake and was drowned. (Luke 8:33)*

Based on this admittedly limited example, it would seem that animals are incapable of tolerating possession by demons. Humans possessed by demons fare little better in Biblical accounts, with the person in question always exhibiting signs that they are tortured in soul and behaving in decidedly abnormal and self-destructive ways. But in Genesis Chapter 3, the serpent appears to suffer no ill effects at being possessed by Satan. There is no struggle, no torture. There is no Biblical basis for explaining how Satan's possession of the serpent could have been so uneventful.

Traditionalists often reference the story of Balaam's donkey in Numbers Chapter 22 in support of a talking animal animated by a soul other than the animal's own. However, that story is wholly inapplicable to the serpent inasmuch as it was God, not a demon, enabling the donkey to speak.

Then the LORD opened the donkey's mouth, and it said to Balaam, "What have I done to you to make you beat me these three times?" (Numbers 22:28)

God is gentle and compassionate, and can exert perfect control over any element of His creation. God has the ability to make an animal speak without causing the animal to suffer. Demons, at least by the one Biblical account available, are incapable of this.

Some traditionalists posit a pre-Fall garden in which all animals were intelligent and could speak. But this view has the makings of a fairly tale and regardless lacks any paleontological evidence. It also provides no basis for concluding that the serpent quietly surrendered to being possessed by a demon. Intelligent creatures that were morally pure would surely resist intrusion by an evil entity.

A less literal interpretation of the serpent's dialogue with Eve would retain Eve's perception that she was speaking to an animal, but would make the serpent to be an apparition rather than a material being. Augustine took this approach, describing the serpent as having entered the garden "spiritually and not bodily" (per Louth). However, while angels appear to humans numerous times in the Bible, there is no example of a demon doing so. The story of Jesus' temptation in the wilderness prior to the beginning of His ministry involved interactions with Satan, but the text is ambiguous about whether Jesus saw Satan or not. If He did, we cannot rule out the possibility that it was His divine nature that enabled Him to see the immaterial Satan, and that any other human would have been unable to do so.

An even further step away from a literal interpretation of the serpent as a Satan-possessed animal would treat the serpent as a figurative description of a human being. Obviously this would have required that a third person be in the garden with Adam and Eve. This is admissible under the Revised Origins Paradigm, and could explain why Eve was not surprised at encountering a speaking "animal." But it would be at odds with the majority of other uses of the Hebrew word for serpent *(nachash*) which identify it as a snakelike animal as described in Chapter 1. Nevertheless, "serpent" is used in at least one place to refer to a human being.

Do not rejoice, all you Philistines, that the rod that struck you is broken; from the root of that snake [nachash] will spring up a viper, its fruit will be a darting, venomous serpent. (Isaiah 14:29, brackets added)

Having a human being labeled as a serpent would also be similar to Jesus referring to king Herod as a fox (Luke 13:32) which was meant to convey Jesus' opinion of Herod's character. Thus the use of "serpent" in Genesis Chapter 3 could have been meant to convey Satan's cunning in trapping a person in the same way that an animal might be a cunning hunter.

Since the portrayal of the serpent in Genesis appears to be unique compared to descriptions of demonic influences elsewhere in the Bible, a rather different approach would be to impose what we know about demons elsewhere in the Bible onto the serpent. That is, Satan's temptation of Eve did not occur through an animal, an apparition, or a person, but rather through the more common mode of thoughts and ideas imposed upon her. There are clear indications elsewhere in the Bible that demons can transmit complex ideas to humans.

The Spirit clearly says that in later times some will abandon the faith and follow deceiving spirits and things taught by demons. (1 Timothy 4:1)

Such "wisdom" does not come down from heaven but is earthly, unspiritual, demonic. (James 3:15)

This approach would require that we treat the serpent and its identification as a wild animal as thoroughly figurative and not at all descriptive of what actually occurred, apart from the fact that there was a dialogue between Satan and Eve.

All of the possible modes of interaction between the serpent and Eve are problematic in some way, and equally so under both the Traditional and Revised Origins Paradigms. But the mode of interaction may be inconsequential. What matters is that there was an interaction, and that that interaction resulted in Eve being deceived into eating from the tree of knowledge.

A more important question is: What was Satan doing in the garden in the first place? The Traditional Origins Paradigm asserts that Satan's presence was part of the same test represented by the tree of the knowledge of good and evil. As discussed above, traditionalists argue that free will is not truly free if there are no alternatives to fidelity to God, and so God was compelled to make some evil alternatives available to Adam and Eve. This view presumes that Adam and Eve would not have sinned if it were not for Satan tempting them to do so. But Satan himself sinned by rebelling against God without having been tempted by any other evil entity, so there is no reason to think that Adam and Eve could not have done with same without being tempted by Satan. Moreover, the same problems generated by the traditional approach to the tree of knowledge as discussed above also apply to Satan's presence in the garden. It amounts to God creating temptations, and at least passively acting in the role of tempter.

There is one other Bible passage that may provide an alternative reason for Satan's presence in the garden of Eden. In Ezekiel's prophecy against the king of Tyre, one portion appears to create a parallel between that king and Satan.

> *You were the seal of perfection, full of wisdom and perfect in beauty. You were in Eden, the garden of God; every precious stone adorned you: carnelian, chrysolite and emerald, topaz, onyx and jasper, lapis lazuli, turquoise and beryl. Your settings and mountings were made of gold; on the day you were created they were prepared. You were anointed as a guardian cherub, for so I ordained you. You were on the holy mount of God; you walked among the fiery stones. You were blameless in your ways from the day you were created till wickedness was found in you. (Ezekiel 28:12b-15)*

This passage describes Satan before his rebellion, and identifies him as having been in the garden of Eden. Since Satan was most certainly not "blameless" in his ways when he tempted Eve, we can deduce that he was in the garden before Adam and Eve arrived there and before his rebellion against God. And being in the garden, he surely had access to the rest of the earth as well. Indeed, it may be that Ezekiel's statement that Satan was "on the holy mount of God" meant that he was also at Horeb where Moses met God in the burning bush.

The earth plays a central role in the life of those in heaven, and we can surmise that this was true all the way back to its initial creation. It would be reasonable to assume that heavenly creatures had access to and roamed the earth long before Adam and Eve existed, and even before any humans existed. When Satan rebelled against God, he was cast out of heaven but retained his access to earth.

> *How you have fallen from heaven, morning star [Lucifer], son of the dawn! You have been cast down to the earth, you who once laid low the nations! (Isaiah 14:12, brackets added)*

Again, for reasons that we cannot fathom, God did not immediately imprison Satan. But His treatment of Satan is essentially no different than His treatment of wicked humans. God continues to allow them the freedom to do as they please for a time.

At the time of the events in the garden of Eden, the world was universally polytheistic. One reason for this state of affairs is that all modern humans have immortal spirits and thus instinctively know that the supernatural world exists, but without guidance from the true God they are inclined to formulate their own ideas of what that supernatural world looks like. Prior to Adam and Eve, they typically did this by identifying a supernatural entity with individual aspects of nature (for example, the sun and the weather) and of human affairs (for example, love and farming). This also enabled them to make some sense out of a universe which seemed chaotic and inexplicable. But another reason for polytheism may be that Satan and his fellow demons were at work to divert humans' attention away from the one true God by establishing rival gods. Satan may even have anticipated that God would reveal Himself to humanity in some way, and wanted to undermine that work.

From this perspective, then, Satan's presence in the garden of Eden was not orchestrated by God to create a temptation for Eve. Instead, Satan was in the garden because he had always been there. And when he sowed mistrust of God in Eve, he was merely continuing the deceptive activity that he had been doing among humanity for hundreds and perhaps thousands of years before.

The sin of Adam and Eve

The Fall, as it was first labeled no earlier than the late 3rd Century, was a choice by Adam and Eve to do what was expressly forbidden by God. There is no disagreement between the Traditional and Revised Origins Paradigms on this point. What motivated them to do so, and the nature of the sin they committed, however, is rather different under the Revised Origins Paradigm. Since that first sin is presumed to be not only the first sin but also the archetype of all subsequent sins for all humanity, it is an important element of the concept of original sin as discussed in Part III.

Under the Traditional Origins Paradigm, the primary sin that Adam and Eve committed in eating from the tree of knowledge was pride. Their motivation was a desire to be independent from God, to be masters of their own fate, and to exercise moral autonomy. While it is the predominant view among early theologians, it is based on a limited reading of Genesis Chapter 3. The serpent tells Eve that eating the fruit will make her like God (Genesis 3:5). Traditionalists interpret Eve's act of eating the fruit as having been motivated primarily or only by this statement. With this as the motivation, they then conclude that Eve wanted to supplant God as ruler of her own destiny.

There is no reason to think that pride was not at work in Adam and Eve, and its role is included in the Revised Origins Paradigm. But a more holistic reading of Genesis Chapter 3 suggests that pride was not the primary motivation. The explanation for Eve's act of eating the fruit is given explicitly, and pride is not obvious in that explanation.

> *When the woman saw that the fruit of the tree was good for food and pleasing to the eye, and also desirable for gaining wisdom, she took some and ate it. (Genesis 3:6a)*

Prompted by the serpent's words, Eve focused her attention on the fruit of the tree of knowledge in a way that she had not previously. She first saw that the fruit was both pretty to look at and could sate her hunger. In this we do not see pride at work, but rather distraction and an indulgent fantasizing about the pleasures of eating the fruit. Diadochus of Photice noticed this process at work:

> *So long as she did not look with longing on the forbidden tree, she was able to keep God's commandment carefully in mind. (per Louth)*

In at least this respect, Eve's actions are consistent with James's description of the source of sin.

> *but each person is tempted when they are dragged away by their own evil desire and enticed. Then, after desire has conceived, it gives birth to sin; and sin, when it is full-grown, gives birth to death. (James 1:14-15)*

Eve took her eyes off God and off His commandment not to eat from the tree of knowledge, and let herself dwell on the value of the fruit that she had previously ignored.

The second part of the explanation for Eve's sin in Genesis 3:6 has been given prominence in most traditional readings: "the woman saw...that the tree was to be desired to make one wise..." Although it is similar to the serpent's statement that eating the fruit would make her like God, it is not exactly the same. Eve focuses on the wisdom she would gain. She may have thought that doing so would make her like God, and that becoming like God would give her independence from God, but the text does not say this. At most, the text indicates that Eve thought she could become wise just as God is wise. As discussed previously, the tree of knowledge offered something that was to be desired, and elsewhere in the Bible wisdom is always characterized as something worth pursuing.

> *Get wisdom, get understanding; do not forget my words or turn away from them. Do not forsake wisdom, and she will protect you; love her, and she will watch over you. The beginning of wisdom is this: Get wisdom. Though it cost all you have, get understanding. (Proverbs 4:5-7)*

Again, this is not to rule out the possibility that pride played a role in Eve's actions. Pride can be inferred from the simple fact that Eve was acting out of self-interest. But it is also clear that there was more going on in Eve than just pride.

A more straightforward reading of Genesis 3:6 is that Eve saw nothing but benefits from eating from the tree of knowledge. There is no mention in her deliberation of the warning God had given about death, nor indeed any mention of God at all. This absence suggests that the serpent had successfully deceived Eve into thinking that God's words were not trustworthy, and as a result God's warning took a backseat to the perceived benefits she would gain from eating the fruit. Paul points out that the serpent's deception was aimed at Eve:

> *For Adam was formed first, then Eve. And Adam was not the one deceived; it was the woman who was deceived and became a sinner. (1 Timothy 2:13-14)*

While traditionalists hold that Eve was deceived into believing that she could become so much like God that she would no longer need God, the text more strongly suggests Eve was deceived into believing that God's words could not be trusted. Why would God deny her something that was so obviously good? Doubt in God's beneficence led to mistrusting Him, and her eyes drifted from God's commandment to her own assessment of the situation. Thus the core failure of Eve would appear to have been mistrust rather than pride.

That said, Eve could not have acted to eat the fruit from the tree of knowledge without believing that, in light of God's untrustworthiness, she was in a better position to judge what was best for her. This, then, was the role of pride in Eve's sin. Faced with the choice of eating something wonderful or being obedient to what might have seemed an unfair and baseless command, she chose to trust her own instincts. Mistrust in God led to greater trust in herself, and an indulgence in her own desires sealed her fate.

Mistrust of God, and more broadly the elevation of trust in oneself over all others, is a core attribute of evolved man. As will be discussed in more detail in Chapter 13, self-centeredness is the engine that drove biological evolution forward. Every creature that has ever existed has placed a priority on preservation and advancement of self, and those creatures who were most successful at doing so lived long enough to pass their DNA on to the next generation. As products of this process, all humans retain this self-centeredness as an inherent characteristic of their being. In mistrusting God, Eve was doing what she was programmed to do.

While Eve's motivation in eating from the tree of knowledge is largely straightforward, Adam's is not. The whole of Adam's role is described in disappointingly few words:

> *She also gave some to her husband, who was with her, and he ate it. (Genesis 3:6b)*

There is no record of a verbal exchange between Eve and Adam about the fruit from the tree of knowledge. However, we know that one took place since, after the Fall, God begins his curse on Adam with this:

> *To Adam he said, "Because you listened to your wife and ate fruit from the tree about which I commanded you, 'You must not eat from it... (Genesis 3:17a).*

Eve might have lied to Adam, manipulated him, begged him, or screamed at him. All we know for certain is that Adam took the fruit that Eve offered him and ate it. The fact that Adam was "with her" may or may not indicate that Adam was present for the entire exchange between Eve and the serpent. He might have been "with her" only during a period of time after that exchange had ended but while Eve was still deliberating about whether to eat the fruit.

Traditionalists fill in the absence of detail about Adam's motives with a narrative that is built around Paul's reflections on the event in the garden. For instance, Paul says that "Adam was not the one deceived" (1 Timothy 2:14) which could be taken as revelatory truth inspired by the Holy Spirit, or could instead be merely Paul's observation that there is no indication in Genesis Chapter 3 that Adam was deceived. Paul also places greater blame on Adam than on Eve, focusing on the implications of Adam's role for the rest of humanity (1 Corinthians 15:20-22 and 45-49, Romans 5:12-19). But it is not clear if Paul viewed Adam as having had a greater awareness of the wrongfulness of his actions, or instead merely saw Adam as the individual responsible to God for the actions of both he and Eve.

The only other source of information we have about Adam's motivation is his own description of his actions.

> *The woman you put here with me - she gave me some fruit from the tree, and I ate it. (Genesis 3:12)*

Here it is at least implied that Adam had given some level of deference to Eve over God, but his motivation is unknown. In short, the Bible is ambiguous about the specific nature of Adam's sin. Nevertheless, traditionalists ascribe the sin of pride to Adam in addition to Eve inasmuch as there is no indication that he hesitated, debated with Eve, or even deliberated within himself. He simply took the fruit and ate it while being fully aware of the wrongness of the deed. With both Adam and Eve charged with the sin of pride, pride became the core sin underlying all other sins and all ills afflicting humanity under the Traditional Origins Paradigm.

But early theologians were not unified in viewing pride as the first or even the worst sin, particularly before Augustine in the late 4th and early 5th Centuries CE. Clement of Alexandria identified the fundamental sin of Adam and Eve as lust, with the fruit of the tree of knowledge being symbolic of their indulging in the pleasures of sexual intercourse before God granted them permission. Clement's contemporary Irenaeus of Lyons thought that Adam's first sin was one of mere thoughtlessness and did not attach a high degree of guilt or culpability to that sin. Others took a similar view according to Bouteneff.

> *The earliest of the Christian authors, like their Jewish predecessors, continued to reckon Adam's sin as mild in comparison to Cain's and other subsequent human fails, but he was still the first.*

There is no reason that pride could not have been working alongside mistrust, and we cannot rule out the possibility that Adam and Eve resented God's command or feared that He had not or would not provide all that they needed. What is clear is that the traditional picture of a self-absorbed duo bent on moral autonomy and a complete rejection of their relationship with God is a considerable stretch of the relevant Bible passages.

They knew they were naked

Under the Revised Origins Paradigm, Adam and Eve had travelled to the garden of Eden from somewhere in Sumer, possibly the city of Eridu. As described in Chapter 6, Eridu had a thriving economy based primarily around fishing, but accompanied by agriculture, animal husbandry, and trade with surrounding cities. There was also a religion based around the patron god Enki. It was a sophisticated civilization where clothing would have been commonplace for protection, adornment, and for religious purposes. Indeed, the human depictions that have been found on artifacts in the region have typically been clothed in some fashion. If Adam and Eve were originally from Eridu, it is highly likely that they would have worn clothing as regularly as we do.

Yet the Bible indicates that in the garden Adam and Eve were naked.

> *Adam and his wife were both naked, and they felt no shame. (Genesis 2:25)*

The Traditional Origins Paradigm holds this statement to mean that Adam and Eve's physical nakedness reflected their moral perfection, innocence, and purity. The difficulty with this view is that there is no record anywhere in the Bible of any resident of heaven, be it God Himself, angels, or departed humans, who are naked. If nudity was a true sign of innocence and purity, surely residents of heaven would likewise be naked. Either Adam and Eve's physical nakedness should not be taken as entirely literal, or it was not meant to indicate that they were naked at all times and in all circumstances.

Some early theologians took Adam and Eve's nakedness as evidence that they were children. As parents know, young children have no shame about nudity, just as the Bible says of Adam and Eve. But it seems unlikely that children would have made the trek from

Eridu to the garden by themselves, though of course with God all things are possible. If they were indeed children, there would be some consistency with Clement's view that their fundamental sin was engaging in sexual intercourse before God had given them leave to do so. In this view they would have grown from children to adolescents over several years in the garden, and increases in hormones would have induced the first sexual desires. Covering their privates with sown fig leaves after the Fall could then be seen as a direct response to their awareness of their sexual sin.

Even if Adam and Eve were adults when they arrived in the garden, their nakedness may nevertheless have been associated with sexuality. The statement that Adam and Eve were naked comes immediately after an editorial statement regarding marriage.

> *That is why a man leaves his father and mother and is united to his wife, and they become one flesh. Adam and his wife were both naked, and they felt no shame. (Genesis 2:24-25)*

Being "one flesh" is indicative not only of unity, but also of sex.

> *Do you not know that he who unites himself with a prostitute is one with her in body? For it is said, "The two will become one flesh." (1 Corinthians 6:16)*

The fact that the reference to their nudity comes immediately after the reference to their being one flesh may indicate not that Adam and Eve were naked at all times, but rather that they were naked at particular times, i.e. when being sexually intimate with one another, and that they had no shame about their sexuality. If so, then the statement that they felt no shame in being naked may have been meant to provide a contrast with the more reserved views of sexuality which existed at the time that the book of Genesis was written.

Under the Traditional Origins Paradigm, Adam and Eve's efforts to cover themselves after their disobedience is both literal and metaphorical: they covered their physical nakedness after gaining an awareness that nakedness was shameful, but they were also attempting to cover the shame of their sin. However, there is nothing in the Genesis story to support the metaphorical interpretation. The text does not say that they became aware that they had sinned. Instead, it merely says that they became aware that they were naked and sought to cover that nakedness.

> *Then the eyes of both of them were opened, and they realized they were naked; so they sewed fig leaves together and made coverings for themselves. (Genesis 3:7)*

Even later when they hide from God, their actions are solely linked to their recognition that they were naked, not to any awareness that they had sinned.

> *He answered, "I heard you in the garden, and I was afraid because I was naked; so I hid." (Genesis 3:10)*

It is God Himself who raises the issue of their disobedience, at which point Adam and Eve's attention shifts from their nakedness to making excuses for their behavior: Adam says it was Eve's fault, and Eve says it was the serpent's fault.

Although Genesis Chapter 3 provides no evidence that Adam and Eve's newfound awareness of their nakedness is a metaphor for an awareness of their sin, in other contexts nakedness is indeed treated this way.

> *They will leave you stark naked, and the shame of your prostitution will be exposed. (Ezekiel 23:29b)*

> *I counsel you to buy from me ... white clothes to wear, so you can cover your shameful nakedness (Revelation 3:18a)*

> *Look, I come like a thief! Blessed is the one who stays awake and remains clothed, so as not to go naked and be shamefully exposed. (Revelation 16:15)*

The context of these passages indicate that the nakedness being addressed is metaphorical, not literal. Specifically, nakedness represents the baring of one's sin. The description of Adam and Eve's nakedness is just as clearly meant to indicate physical nakedness. It may be that there is an additional metaphorical meaning to their nakedness that is unstated in Genesis, but at a minimum it appears clear that their nakedness must be treated as literal.

It is important to note that Adam and Eve were not completely naked when they hid themselves from God even though Adam referred to himself as being naked. In Genesis 3:7 we are told that they sewed fig leaves together to make themselves coverings, and there is no reason to believe that they were not wearing those coverings when they later hid themselves from God. Thus when Adam told God that he was afraid because he was naked, he was not referring to being completely naked. Rather, he was saying that he was not sufficiently clothed or was in some way inappropriately clothed.

There is evidence elsewhere in the Bible of a particular decorum or modesty expected of God's rational creatures when they are in His presence. God gives very specific instructions to Moses about the type of clothing that the priests must wear when serving in the Tabernacle (Exodus 28:4). John makes a point of describing the clothing of certain angels as fine linen that was pure, white, and bright (Revelation 15:6 and 19:14), and in one place even describes an angel whose only clothing appears to be a cloud (Revelation 10:1). But two particular descriptions stand out.

> *Above him were seraphim, each with six wings: With two wings they covered their faces, with two they covered their feet, and with two they were flying. (Isaiah 6:2)*

> *They [cherubim] each had two wings spreading out upward, each wing touching that of the creature on either side; and each had two other wings covering its body. (Ezekiel 1:11, brackets added)*

Neither passage gives a reason for why the angels use their wings to cover their bodies, but commentators speculate that it is some combination of humility, an expression of reverence, a sense of unworthiness in the presence of pure holiness, and a recognition of their imperfection in comparison to God's perfection. Whatever the reason, it would be reasonable to assume that Adam and Eve had gained a similar perspective through eating from the tree of knowledge, and similarly sought to cover their bodies as a result.

Whatever clothing they wore upon their arrival in the garden, which heretofore had seemed appropriate, was no longer so.

The coverings that Adam and Eve sewed together from fig leaves have traditionally been seen as coverings for genitals, and Adam and Eve's efforts to cover themselves a sign that they were previously wearing no coverings at all over their genitals. The Hebrew word *chagowr* which is translated as coverings in Genesis 3:7 in the New International Version is translated as loincloths, apron-like girdles, and aprons in other versions. In other parts of the Old Testament, it seems clear that *chagowr* refers to something worn about the waist.

> *Instead of fragrance there will be a stench; instead of a sash [chagowr], a rope (Isaiah 3:24a, brackets added)*

> *Joab said to the man who had told him this, "What! You saw him? Why didn't you strike him to the ground right there? Then I would have had to give you ten shekels of silver and a warrior's belt [chagowr]." (2 Samuel 18:11, brackets added)*

> *He killed them, shedding their blood in peacetime as if in battle, and with that blood he stained the belt [chagowr] around his waist and the sandals on his feet. (1 Kings 2:5b, brackets added)*

What is not clear is whether the only or primary purpose of a *chagowr* is to cover genitals, or instead is to cover hips, thighs, or legs. Given the ambiguity, an admissible interpretation of Adam and Eve's efforts to create clothing from fig leaves is that they were attempting to cover more of themselves than was already covered. That is, their genitals were already covered, but they had become aware of the inappropriateness of waists, thighs, or legs remaining uncovered.

As something worn about the waist, a *chagowr* does not appear to be something that covers the chest. As a result, it would not have covered Eve's breasts, and there is no mention of Eve attempting to cover her breasts in her efforts to cover her nakedness. While it is possible that Eve did not consider breasts as a body part that warranted covering, it seems more likely that Eve's breasts were already covered and thus needed no additional attention.

In summary, the story of Adam and Eve's nakedness is not inconsistent with a pair of travelers from Eridu who were already clothed to some degree, but who came to an awareness of the necessity of propriety in attire after eating from the tree of knowledge.

Conclusion

Adam and Eve's time in the garden was meant to be preparatory, and thus temporary. Having come from a polytheistic culture, they needed to grow in their understanding of the one true God, His ways, and His plans before they could be the emissaries to the rest of humanity that they were called to be. The tree of the knowledge of good and evil was in the garden not as a setup to test them, but as an necessary element in the process of maturation that would come into play at the proper time, if only they would trust fully in God. Prompted by Satan, they instead mistrusted God and ate from it before being given permission to do so.

Chapter 9
The Curse

In response to the disobedience of Adam and Eve, God meted out punishments to both of them, as well as to the serpent. Their life of comfort was over and their access to the tree of life lost. Henceforth their existence would be marked by striving and strife, and would end in death.

The Traditional Origins Paradigm holds that the Fall was much more than just a failure of Adam and Eve to keep a simple commandment. It was the beginning and source of a new state of being for humanity, the hopeless sinner. Every single human after Adam and Eve would be tainted by a corrupt and inescapably sinful nature, something generated by Adam's disobedience and passed down to his children and thus all humanity. And being in such a hopeless state, every single human being is in need of a Savior to deliver him from his sins.

Under the Revised Origins Paradigm, it remains true that the sin of Adam and Eve had dire consequences both for themselves and for all of mankind. However, the manner in which those consequences propagated to the rest of the world is rather different. Much of this will be discussed in Part III. This chapter focuses on the nature and purpose of God's punishments in light of His relentless efforts to bring all of humanity into relationship with Himself.

God punishes the serpent

Once Adam and Eve's disobedience is exposed, God begins the declaration of punishments with the serpent. Rather paradoxically, He curses the serpent whom tradition holds was already cursed.

> *So the LORD God said to the serpent, "Because you have done this, cursed are you above all livestock and all wild animals! You will crawl on your belly and you will eat dust all the days of your life. (Genesis 3:14)*

Satan's original rebellion against God is presumed to have occurred before Adam and Eve arrived in the garden, and possibly before time in our universe began; there is no tradition wherein Satan's original rebellion is the same as or coincident with his act to deceive Eve. Moreover, Satan is presumed to have been cursed at the time of his original rebellion, before God's curse on him in the garden. In Jesus's parable of the sheep and goats, He refers to the fact that hell was created for the purpose of punishing Satan, a curse that is generally understood to have occurred before the events of the garden.

> *Then he will say to those on his left, "Depart from me, you who are cursed, into the eternal fire prepared for the devil and his angels. (Matthew 25:41)*

But while Satan's ultimate punishment is fixed, it is delayed. For reasons that only God knows, God did not send him into the eternal fire immediately but instead gave him considerable freedom, albeit with restrictions.

If God's curse on the serpent was meant to impose new punishments that were precipitated by his deception of Eve, it is unclear how those new punishments differed from or otherwise added to the curse that Satan was already under. One can deduce a sense among early theologians that the curse was intended to restrict Satan's freedom to move and act in some way, but there is essentially no discussion of what Satan could do before the garden curse that he could not do afterwards. In fact, there is a case to be made that Satan actually has more freedom after the Fall inasmuch as Adam and Eve effectively gave their authority to rule the earth over to him.

> *The devil led him up to a high place and showed him in an instant all the kingdoms of the world. And he said to him, "I will give you all their authority and splendor; it has been given to me, and I can give it to anyone I want to. (Luke 4:5-6)*

Given the absence of any other Biblical support for a two-stage curse on Satan - one resulting from his original rebellion and another in the garden - it seems more likely that God's curse on Satan in the garden was not in fact something new but rather was a reiteration of the curse that Satan was already under, spoken aloud for the benefit of Adam and Eve and us.

However, God's curse on the serpent did contain an element that was new. God says that there will be strife between the (presumably spiritual) offspring of the serpent and those of Eve.

> *And I will put enmity between you and the woman, and between your offspring and hers; he will crush your head, and you will strike his heel. (Genesis 3:15)*

This verse is universally understood to be prophetically Messianic: Christ will utterly defeat Satan, but in the process Satan will wound Christ. But this verse also implies something else. Had Adam and Eve not been disobedient, there would have been no such enmity. The dominion that Adam and Eve had been given over all the animals of the earth would have extended to the serpent as well. Satan would have been cast out of the garden, and possibly out of the whole earth. Again, it would appear that Adam and Eve's disobedience gave Satan the right to remain on earth, and to more or less do as he pleases. The book of Job makes a reference to this freedom.

> *The LORD said to Satan, "Where have you come from?" Satan answered the LORD, "From roaming throughout the earth, going back and forth on it." (Job 1:7)*

As described in Chapter 8, Satan had begun roaming the earth long before Adam and Eve existed, and before he rebelled against God. At the time of the events of the book of Job, Satan was doing what he had been doing since that rebellion: roaming the earth and wreaking havoc on God's children. This was one of the disastrous consequences of Adam and Eve's sin.

God curses the ground

God's punishment for Adam is somewhat indirect: the ground is cursed rather than Adam himself, and the pain that Adam will experience will be a result of the extra effort that will be needed to cultivate sufficient edible plants to sustain he and Eve.

> *Cursed is the ground because of you; through painful toil you will eat food from it all the days of your life. It will produce thorns and thistles for you, and you will eat the plants of the field. (Genesis 3:17b-18)*

Under the Traditional Origins Paradigm, God's curse on the ground applies to the entire earth, the garden included. The garden would change from being a lush, well-watered region replete with many edible plants to a dry wasteland overwhelmed by thorns and thistles. However, this passage need not be interpreted this way. It may be that the reference to thorns and thistles, and the difficultly Adam would have in cultivating sufficient edible plants to eat, were references to the world outside of the garden. As described in Chapter 2, the garden was particularly well suited for humans while the world outside of the garden was less hospitable.

One indication that the garden itself was not included in God's curse is that God sends Adam and Eve out of it and takes steps to ensure that they cannot return to it.

> *"He [Adam] must not be allowed to reach out his hand and take also from the tree of life and eat, and live forever." So the LORD God banished him from the Garden of Eden to work the ground from which he had been taken. After he drove the man out, he placed on the east side of the Garden of Eden cherubim and a flaming sword flashing back and forth to guard the way to the tree of life. (Genesis 3:22b-24, brackets added)*

The primary reason given for the banishment was to prevent Adam from eating from the tree of life. But if the garden were to become a wasteland along with all the rest of the earth, the tree of life would surely wither along with all other plants, and there would have been no need to prevent access to it. The implication of keeping Adam and

Eve out of the garden is that the tree of life would continue to flourish and that its fruit would remain fresh, just as it would be for all other plants within the garden. The garden would remain a garden. Outside the garden is where the thorns and thistles would grow. Had already been growing, in fact.

God's curse on the ground was a declaration that all hope of the garden expanding to fill the earth was now gone. The ground outside the garden would remain as it was, a wilderness, and Adam and Eve would now be consigned to live in that wilderness.

God punishes Adam and Eve

As for God's curse on the serpent and on the ground, God's punishments for Adam and Eve were also mixtures of new punishments and declarations of what was already true. Adam's punishments can be described as falling broadly into two categories. One is the punishment of death.

> *By the sweat of your brow you will eat your food until you return to the ground, since from it you were taken; for dust you are and to dust you will return. (Genesis 3:19)*

This punishment was addressed in more detail in Chapter 5. In short, God's proclamation of death after the Fall was not the first appearance of death on the earth, but rather a removal of the promise of eternal life that was Adam's so long as he remained faithful to God. Death was not new, but its application to Adam was.

The other punishment on Adam concerns the additional effort it would require for him to work the ground.

> *through painful toil you will eat food from it all the days of your life... By the sweat of your brow you will eat your food (Genesis 3:17b, 19a)*

There is no doubt that God's punishment meant that Adam would suffer in ways that he had not prior to the Fall. Specifically, it would be more difficult for Adam to cultivate the ground in order to produce sufficient food to eat. His punishments were new in that sense. What was not new was that the ground outside the garden was already less well suited for agriculture. Adam would be sent out of the garden, and it is there that he would find work toilsome.

Eve's punishment was likewise twofold. The first related to the pain of childbirth.

> *I will make your pains in childbearing very severe; with painful labor you will give birth to children. (Genesis 3:16a)*

Assuming that the pain in question was physical rather than emotional pain, a concept discussed in more detail in Chapter 4, this punishment need not refer to the creation of pain where none existed before. Instead, it may refer to an increase in pain from some to more. In the same way that the fruit from the tree of life conferred general health and longevity, it may also have minimized the pain of childbirth. Without access to the tree of life, the analgesic benefits of that fruit would be gone. Pain itself was not new, but its presence in childbirth for Eve was.

The other punishment on Eve concerned her relationship with Adam.

> *Your desire will be for your husband, and he will rule over you. (Genesis 3:16b)*

God is clearly describing a change in the relationship between Adam and Eve as a result of the Fall, but the precise nature of that change is not clear. Surely Eve desired her husband before the Fall, so the first phrase may mean that her desire for Adam would surpass his desire for her, resulting in a mismatch in the intensity of their mutual affection for one another. If this is the case, then the second phrase may be describing a complementary change in Adam's attitude towards Eve, one less of respect and care and more of control.

The traditional view of the relationship between Adam and Eve before the Fall is that God made Adam the head of the marriage and the head of the family. Eve was to "help" Adam in an essentially subordinate role. Paul appears to place the origin of these marriage roles in the pre-Fall creation story, not in the punishments after the Fall.

> *A man ought not to cover his head, since he is the image and glory of God; but woman is the glory of man. For man did not come from woman, but woman from man; neither was man created for woman, but woman for man. (1 Corinthians 11:7-9)*

> *I do not permit a woman to teach or to assume authority over a man; she must be quiet. For Adam was formed first, then Eve. (1 Timothy 2:12-13)*

As a result, the phrase "he will rule over you" in Eve's punishment after the Fall cannot mean that God was establishing a new role for Adam as head of the marriage, as Adam was already head of the marriage. But the use of the word "rule" to describe that role is new. It may be that Adam would approach his role more earnestly but less responsibly after the Fall, exercising it in a more domineering fashion. He may also have become less in tune with Eve's emotional needs and less able to reciprocate. Their relationship, it seems, would never again be the synchronous partnership that God had intended it to be.

Notably absent from the punishments on Adam and Eve are the sorts of punishments we see God impose on later generations when they turned from Him. God does not decree sickness or blindness or misfortune. He does not send wild animals to attack them, nor pestilence or other disaster. The punishments on Adam and Eve are bad, but not nearly as bad as they could have been. Moreover, their punishments appear to be more consequential than imposed. That is, they are connected to other aspects of normal life. For Eve, the pain she would experience would be the result of childbirth. For Adam, the pain he would experience would be the result of the toil required to gather sufficient food. God was not merely imposing pain or causeless suffering, but rather giving them over to a life that would have none of the benefits of living in the garden and having a close fellowship with Himself. God's special blessings and protection would be removed, and Adam and Eve would experience what life was like without them.

God's treatment of Adam and Eve is also reflected in the fact that, as pointed out earlier, there is no curse on them; the word "curse" is only used in reference to God's punishments on the serpent and on the ground. As curses represent eternal disfavor and rejection with no hope of relief or redemption, we can conclude that God's punishments for Adam and Eve, however painful, were meant to be disciplinary rather than condemnatory. And when God disciplines, He remains in relationship with those He is disciplining.

> *because the Lord disciplines the one he loves, and he chastens everyone he accepts as his son. (Hebrews 12:6)*

This conclusion is rather at odds with one of the tenets of the Traditional Origins Paradigm, that the Fall resulted in an unqualified rift between God and man, effectively an end to the relationship. It is sometimes depicted as an uncrossable chasm that only Christ can bridge. While it is certain that Christ provided the means through which our relationship with God is fully restored, it is not the case that the relationship between humanity and God ended after the Fall. Far from it. After pronouncing His judgments, God immediately does something distinctly parental: He makes garments to cover their nakedness. After they leave the garden, humanity continues to have a relationship with God, faulty though it may have been. Cain and Abel bring offerings to God, and God speaks directly to Cain after he kills Abel. When Eve bears Seth, she credits God. The rest of the Old Testament, the story before Jesus arrives on the scene, is entirely about God's efforts to reach out to humanity and to draw them back to Himself. The Fall damaged the relationship between God and humanity but did not destroy it. Or, more precisely, the relationship that God had intended to create between Himself and His children through His special treatment of Adam and Eve never materialized.

The expulsion from the garden

The traditional reading of the expulsion from the garden is that God was following through on His warning from Genesis 2:17 that anyone who ate from the tree of knowledge would die. But similar to the other punishments on Adam and Eve, the punishment of death was indirect.

> *Then the LORD God said, "Behold, the man has become like one of us in knowing good and evil. Now, lest he reach out his hand and take also of the tree of life and eat, and live forever -" therefore the LORD God sent him out from the garden of Eden to work the ground from which he was taken. (Genesis 3:22-23, ESV)*

God did not directly impose death but instead eliminated the possibility of immortality.

The Revised Origins Paradigm also understands this passage as describing the death of Adam and Eve, but not as a fulfillment of God's warning in Genesis 2:17. God had already declared several verses earlier that they would die as a result of their disobedience, and there would have been no need to repeat it. Instead, this passage describes the means through which death would occur, and provides this description not for the benefit of Adam and Eve but rather for the benefit of the host of heaven. God speaks of Adam in the third person, and directs His comments to "us." While it is possible that "us" could refer to the plurality of the Trinity, the interjection "Behold" is meant to direct the attention of others to the new state of affairs. As a result, it is more likely that "us" is a reference to all the rational creatures of heaven. This is consistent with the fact that in the very next verse He assigns angels to cut off access to the tree of life.

> *After he drove the man out, he placed on the east side of the Garden of Eden cherubim and a flaming sword flashing back and forth to guard the way to the tree of life. (Genesis 3:24)*

Regarding the reason that Adam and Eve were banished from the garden, it is commonly held that it is because they gained knowledge by eating from the tree of the knowledge of good and evil and thereby became like God. This is the result of treating "Now" in Genesis 3:22 as "Therefore." But not only are there no translations which use "Therefore" instead of "Now," as described in Chapter 8 there is no reason to believe that the knowledge Adam and Eve gained was anything but good, and thus no reason for God to decree death for them because of that knowledge (unless we view God as an egocentric overlord intent on preventing others from having what He has). Moreover, if the knowledge that Adam and Eve acquired was indeed the same as God's knowledge and they had become like God, there is no obvious reason why they should be prevented from being immortal as God is immortal. If anything, that knowledge would make immortality more, not less fitting for them. The more likely explanation for their expulsion from the garden is the fact of their disobedience to an unambiguous command.

Even so, there appears to be more to the expulsion from the garden that mere punishment. After all, Adam and Eve would spend many years living and working before they died. What would be God's purpose in allowing them to live such long lives if His true goal was for them to die?

Some of the punishments given to Adam and Eve can be correlated directly to depriving them of the benefits of the tree of life. The traditional view of the tree of life is that its sole purpose was to bestow immortality on anyone who ate its fruit. But as described in more detail in Chapter 5, it is more likely that the tree of life conveyed not just immortality but also exceptional vigor not unlike that given to Isaiah through an angelic offering of food and drink in 1 Kings 19:5-8. By eating of it, Adam and Eve would have experienced extra strength and endurance, freedom from sickness, accelerated healing from injuries, and potentially a muted pain response. By being deprived of access to the tree of life, these special advantages would be lost.

But while some of the punishments can be connected to the tree of life, all of the punishments can be correlated with God's decision to send them out of the garden. It was outside the garden where the ground was "cursed" and where Adam would find his efforts to work it for sustenance toilsome. Whereas the garden was the zone of God's special blessings and protection for His chosen emissaries-in-training, the world outside the garden was a place where nature and the elements had run riot and where inconvenience and difficulty would be common. The greater hardships of life outside the garden, combined with the disharmony that appeared between Adam and Eve during and immediately after the Fall, would lead to greater stress in their relationship. The marriage hierarchy established by God, intended to be a source of order and purpose, would instead become a source of friction. All of these outcomes were described earlier in Genesis Chapter 3 as punishments on Adam and Eve for their disobedience, and all of them are fulfilled in Adam and Eve's expulsion from the garden.

But the means through which God would punish Adam and Eve would be the same means through which He would discipline them and, ultimately, put them back on track. In their new, more difficult life outside the garden, Adam and Eve would experience pervasive discomfort and the threat of lack, and would discover the necessity of trusting God in order to cope. In His mercy, God intentionally and actively subjected Adam and Eve to difficult circumstances in order to elicit the kind of trust that they could not sustain in the garden when circumstances were pleasant.

This is the first of what would be many instances in the Bible of blessings leading to conceit and rejection of God, and God using difficulty to bring His children back into relationship with Himself. Most such stories involve Israel. For instance, God warns the Israelites of the dangers of comfort immediately prior to their entry into the promised land.

> *When the LORD your God brings you into the land he swore to your fathers, to Abraham, Isaac and Jacob, to give you - a land with large, flourishing cities you did not build, houses filled with all kinds of good things you did not provide, wells you did not dig, and vineyards and olive groves you did not plant - then when you eat and are satisfied, be careful that you do not forget the LORD, who brought you out of Egypt, out of the land of slavery. (Deuteronomy 6:11b-12)*

After defeating their enemies and settling into their new homes, and after Moses and Joshua die, the Israelites do exactly what God predicted. The book of Judges abounds with the repeating cycle of blessing-disobedience-punishment-repentance-blessing. Elsewhere the Bible seems to suggest that difficulty is a valuable, and possibly necessary component of our spiritual maturation.

We must go through many hardships to enter the kingdom of God (Acts 14:22b)

Consider it pure joy, my brothers and sisters, whenever you face trials of many kinds, because you know that the testing of your faith produces perseverance. (James 1:2-3)

The events of the garden demonstrate that God would rather use comfort and blessings to draw His children to Himself and to train them into spiritual maturity, but that comfort and blessings are much more likely to impel them to trust themselves more than they trust God. God must then revert to difficulty as His tool for maturation.

The impact of the Fall on creation

In addition to the explicit punishments, the Traditional Origins Paradigm posits that all of earthly creation was also adversely affected by the Fall. Every living thing began aging and death became a reality for all animals and plants. Herbivores became carnivores, disease-causing microbes and poisonous plants came into existence, snakes developed venomous fangs, wasps began laying their eggs in live caterpillars, and tornadoes, floods, and droughts appeared. Traditionalists presume that all of these effects of the Fall on creation are included within God's curse on the ground.

Cursed is the ground because of you; through painful toil you will eat food from it all the days of your life. It will produce thorns and thistles for you (Genesis 3:17b-18a)

But the corruption of creation is an enormous amount of activity to derive from this single Bible verse. Worse, this verse would also need to account for the appearance of accidents and misfortune which, under the Traditional Origins Paradigm, could not have existed before the Fall if there had been no such thing as suffering for man or animal.

Because the list of punishments and curses in Genesis Chapter 3 were meted out by God, by extension God would also have been responsible for the adverse impacts on creation which are presumed to be included but unstated in the verse above. That is, it was He who created thorns and stinging insects and the meat-eating lion after the Fall. But according to Genesis 2:2-3, God ceased from all His creative acts on the sixth day. The author of Hebrews emphasizes this fact:

> *his works have been finished since the creation of the world. For somewhere he has spoken about the seventh day in these words: "On the seventh day God rested from all his works." (Hebrews 4:3b-4)*

Recognizing this obstacle, some traditionalists have proposed a genetic process wherein herbivores were originally created with the capacity to be carnivorous, and edible plants were originally created with the capacity to be poisonous. They were all harmless and beneficial in their original forms despite having the potential to become something more dangerous and destructive, and the Fall provided the necessary trigger to transform them into the forms we know today. Genetically, they remained exactly the same before and after their transformation, and thus no biological evolution was involved. Instead, it was merely a change in the expression of their genes. This same process is also used in the Traditional Origins Paradigm to explain how Noah's ark could have contained all species of animals: the ark contained only common ancestor "kinds" which then diversified to become all known species after they disembarked.

That different expressions of genes can produce different physical traits has been known for some time. Selective breeding and cultivation, sometimes termed artificial selection to distinguish it from natural selection, can produce plants and animals with different colors, sizes, aptitudes, and to some degree forms. But in such cases there is no change in species, only variations within a species. There is no reliable scientific information to support the concept that, for instance, a plant-eating animal could be transformed into a meat-eating one merely through changes in gene expression, or that a single rodent "kind" could diversify into the 1,500 species of rodents that exist today. Instead, the genes themselves would need to change. And if such changes were to occur over anything less than geologic timescales, they would require direct action by God.

Even if one were to allow the possibility that plants and animals were originally created with the capacity to become dangerous and destructive, one is left with the necessary conclusion that God chose to make them so. That is, He effectively prepared for the Fall before it occurred. This would make Him at least complicit in the arrival of suffering and death on the earth. While it would remain true that the trigger which released suffering and death was Adam and Eve's sin, nevertheless if God had not instilled the capacity for predation and poison within creation, it could not have been released.

The awkward position into which God is placed in this approach is not unlike the one that traditionalists decry in the context of biological evolution. There, traditionalists contend that biological evolution, with its focus on natural selection and survival of the fittest, makes God out to be the creator of suffering and death, something that they contend cannot be true of a good God. Suffering and death were addressed in Chapters 4 and 5, respectively, where it was shown that their existence is in fact not inconsistent with a good God whose ways are inscrutable.

To remove God entirely from the corruption of creation, some traditionalists concede that genetic changes would have been needed, but argue that those changes would have occurred entirely as a result of the release of sin on the earth. This approach admits to limited forms of biological evolution, sometimes termed microevolution which results from changes in gene frequency within a population. It can be observed over relatively short timescales such as decades and even a single year, such as when bacteria develop resistance to antibiotics and when viruses mutate to create new strains. But the transformation of a herbivorous lion into a carnivorous lion is orders of magnitude more complex, and the two versions of lions would most certainly not be the same species. This is not something that could occur through the very limited process of microevolution.

More problematic in the idea that Adam and Eve's sin caused some form of genetic changes in plants and animals is the absence of any hypothesis for the mechanism through which it would occur. Sin is a condition rather than a substance, and one that applies only to sentient creatures. There is no basis for assuming that it can operate like a contagion, spreading outward from Adam and Eve to infect plants and animals and so modify them at the genetic level in very precise ways. Moreover, such a "sin sickness" could not have chosen which herbivores to change into carnivores and which to leave alone to produce the mix we have today, and could not have whipped wind into tornadoes or caused a volcano to erupt.

One might speculate that Adam and Eve's sin opened the door for Satan to do these things. He is, after all, sentient and capable of influencing the weather (Job 1:19) and directing the action of diseases (Job 2:7). Whether he also had the ability to modify animal and plant genes is unknown. The primary difficulty with this approach is that there is no paleontological evidence to support it. That is, there is no sudden shift observable in the fossil record at around 4000 BCE from a thoroughly herbivorous ecosystem to a mixture of herbivores and carnivores. If Satan was in any way responsible for the rise of carnivores and poisonous plants, it would have had to occur over the millions of years prior to Adam and Eve, and would be indistinguishable from the purely natural process that is termed biological evolution (or macroevolution).

Apart from Genesis 3:17-18, the Bible is essentially silent on what impact Adam and Eve's sin may have had on the rest of creation. One New Testament passage that does address the current state of creation does not provide any obvious connection to Adam and Eve or their sin.

> *For the creation was subjected to frustration, not by its own choice, but by the will of the one who subjected it, in hope that the creation itself will be liberated from its bondage to decay and brought into the freedom and glory of the children of God. We know that the whole creation has been groaning as in the pains of childbirth right up to the present time. (Romans 8:20-22)*

Both the Traditional and the Revised Origins Paradigms agree that at the end of the age, when all God's children come fully into their inheritance and receive the redemption of their bodies (verse 23), so too will the rest of creation.

> *He who was seated on the throne said, "I am making everything new!"" (Revelation 21:5a)*

God will ultimately complete the process of bringing order out of disorder and unifying all things with Himself that He had begun at the beginning of time. The disagreement is in the reason for creation's current state, an issue which the Romans passage above does not address. The Traditional Origins Paradigm interprets the "bondage to decay" as being rooted in God's curse on the ground and, by extension, all of creation. But the passage neither cites that curse nor suggests that creation was originally created perfect and at some time afterward became corrupted. It merely says that creation is currently in bondage to decay. Moreover, the predominant view among theologians is that "him who subjected it" is a reference to God, not Adam or Satan. Thus Romans 8:20-22 can just as easily be read to say that God intended all plants and animals to be mortal from their very first appearance on earth.

Another passage that traditionalists interpret as referencing the impact of the Fall on creation is the following:

> *In the beginning you laid the foundations of the earth, and the heavens are the work of your hands. They will perish, but you remain; they will all wear out like a garment. Like clothing you will change them and they will be discarded. (Psalm 102:25-26)*

Again, traditionalists hold that the reason for creation's subjection to decay is implied but not stated. A more straightforward reading is that it was an inherent attribute of creation from the beginning to be subject to entropy, and thus to be destined to wear out like a garment.

Finally, passages describing death as having entered the world through one man, such as in Romans Chapter 5, are exclusively focused on humanity. The end of the curse described in Revelation is likewise associated with humanity, not the rest of creation.

> *And the leaves of the tree are for the healing of the nations. No longer will there be any curse. (Revelation 22:2b-3a)*

What is healed is the nations, meaning all people. And following this healing, nothing will remain that is accursed.

The Fall affected Adam and Eve and their relationship with God. It did not affect creation in any direct way. It did, however, mean that creation would not benefit from the outward expansion of God's favor and blessing from the garden that might have occurred if Adam and Eve had remained true to their calling. This is the meaning of God's curse on the ground. Creation would instead remain imperfect until the end of time when everything will come under God's headship and everything will be made new.

Conclusion

As is so often the case, God gave Adam and Eve a simple and unambiguous command without providing an explanation for it. He was looking for trust and fidelity in the face of any opposing thought, feeling, or message. When Adam and Eve ultimately broke that command, it was in perfect keeping with His justice that He punish them.

However, it was never God's intention to reject humanity, the apex of all creation. As He would demonstrate time and again with the Israelites, His goal was to repair the connection that had been damaged, and His punishments were largely designed to do this. The only explicit curses that God declared were directed toward the serpent and the ground. The serpent will never be redeemed, and the ground will only be redeemed at the end of the age when all things are made new. But human beings can be redeemed at any time by placing their trust in the God that Adam and Eve did not trust.

Part III

Original Sin

Chapter 10
The Emergence of the Doctrine of Original Sin

The disobedience of Adam and Eve in the garden of Eden was unprecedented and pivotal, and everything that happened between God and humanity after that point has its roots in that seminal event. It is just as important a component of the Revised Origins Paradigm as it is under the Traditional Origins Paradigm.

But even if every other element of the Revised Origins Paradigm can be harmonized with the Genesis story - most particularly, the fact that Adam and Eve were the products of biological evolution and thus not the only humans on earth during their time in the garden - it is incomplete if it does not address the concept of original sin and humanity's universal and unqualified need for redemption through Christ. This requires a deep dive into these critical doctrines and thus into the very heart of Christianity. It is dangerous territory, and bound to make any reasonably mature Christian at least mildly nervous. But it is not unprecedented. Not only has the church taken centuries to arrive at the ultimate formulations of several key doctrines that are now treated as orthodox, but it has modified its views on a number of doctrines at different points in history. In the case of the Revised Origins Paradigm, the departure from the traditional view of original sin represents not so much something new as it does a reversion to views held by many theologians in antiquity.

The problem

If biological evolution is the true process through which humanity came to be, there is essentially no way in which humans can be seen as having been perfect in their pre-Fall state, morally or otherwise.

Evolution is a messy process, fraught with competition and conflict and fueled by self-interest. Every individual instinctively seeks to advance itself at the expense of others. Every species of plant, insect, and animal that has ever existed is therefore imbued with the qualities that made them successful in the game of survival of the fittest. For humans, these qualities are generally at odds with those of the perfect God.

The innate imperfection of man is an unavoidable component of the Revised Origins Paradigm, and is diametrically opposed to the traditional view. Under the Traditional Origins Paradigm, Adam and Eve's pre-Fall perfection and the associated perfection of all creation in its original state is the starting point of the story of humanity, and it is the only way in which the Fall can have any meaning. From whence did man fall if not from a state of perfection?

Just as problematic for traditionalists is the fact that, if Adam and Eve were the products of biological evolution, then they were not the only humans on the earth at the time of the events in the garden of Eden. As a result, even if their act of disobedience did produce in them some new sinful nature that was subsequently passed on to their children, that sinful nature would not be passed on to the children of anybody else. Those persons who were the direct descendants of Adam and Eve would have a sinful nature, while everyone else would not have a sinful nature. Consequently, some would need to be redeemed through the blood of Christ while others would not.

Christ is the center of Christianity, and His free offer of justification for those who believe in His redemptive work on the cross, combined with His actual and our eventual resurrection, is what sets Christianity apart from all other religions. Traditionalists are right to be suspicious of any doctrinal view that appears to call into question the helpless state of every single person before the holy God. It is not surprising, then, that they view biological evolution as wholly and hopelessly inconsistent with Christianity, and they condemn of any consideration of it. Mortenson & Ury put it this way:

> *... the old-earth view of creation gives away the very heart of the Christian apologists' message that Christ came to redeem fallen humanity and the fallen creation.*

Collins casts the issue as one affecting not just Christian doctrine but the philosophy of human existence.

> *If we abandon the conventional way of telling the Christian story, with its components of a good creation marred by the fall, redemption as God's ongoing work to restore the creatures to their proper functioning, and the consummation in which the restoration will be complete and confirmed, then we really give up all chance of understanding the world.*

The atheist G. Richard Bozarth states the situation perhaps even more plainly than any theologian:

> *...evolution destroys utterly and finally the very reason Jesus' earthly life was supposedly made necessary. Destroy Adam and Eve and the original sin, and in the rubble you will find the sorry remains of the son of god. Take away the meaning of his death. If Jesus was not the redeemer who died for our sins, and this is what evolution means, then Christianity is nothing. (per Holmes)*

From the perspective of traditionalists, a truly faithful reading of the Bible can only lead to one doctrinal outcome: the Traditional Origins Paradigm. Any reading that leads to a different conclusion can only mean that the reader is guilty of heresy through eisegesis, choosing for himself what he wants the Bible to mean. If a Revised Origins Paradigm is to have any legitimacy and warrant honest consideration, then, it must demonstrate some minimum level of consistency with long held doctrines of original sin and humanity's universal need for a Savior. Those doctrines cannot remain exactly as they are if a Revised Origins Paradigm is true, but neither is it necessary to alter them so dramatically that they would be unrecognizable to a Christian in the early centuries of the Common Era.

Development of doctrine

There is a strong sense among traditionalists that the collection of doctrines embodied in the Traditional Origins Paradigm has existed since the beginning of Christendom, and has been embraced by all orthodox Christians since that time. For instance, in the context of repudiating secular views of the doctrines of sin and salvation, Boice takes as a given that orthodox Christian doctrine has always existed in its current form.

> *...the theological terms that we have always used and which the church continues to use (because it is part of its heritage) are being redefined.*

In reality, theologians at the beginning of the Common Era debated for decades and sometimes centuries over the specifics of such doctrines as the Trinity, the dual God/man nature of the Christ, baptism, and the Eucharist. The general forms of these doctrines, it is true, can be traced back to the 1st Century, but the precise details were by no means obvious at that time.

It was no different with the doctrine of original sin. It was generally held in the 1st Century that Adam's disobedience in the garden of Eden had disastrous effects for all mankind, and that Jesus came to bring us into the relationship with God that He had always intended us to have. But it would not be until the end of the 4th Century that Augustine of Hippo would begin to formulate the concept of original sin that we know today, and it was not until the Council of Orange in 529 CE that it would be formally and finally accepted by the church as doctrine. There is no basis on which to assert that this doctrine existed in the 1st Century, let alone was accepted by Christians of that time.

This is not to say that Augustine's formulation was wholly new. As for the development of other doctrines, Augustine drew upon ideas that had been developed by his spiritual forebears, particularly Ambrose and Ambrosiaster earlier in the 4th Century. They in turn had expanded and modified ideas developed earlier still. The earliest concept of original sin bearing at least some resemblance to Augustine's was developed by Irenaeus in the 2nd Century. But as described below, Augustine's formulation of the doctrine of original sin was unique in several key respects.

Not only did Augustine's formulation of original sin not appear until the 4th Century, it did not at first gain wide acceptance. He was accused of being an innovator by some, deviating from traditional views on the matter of free will, predestination, and irresistible grace. Contemporaries of Augustine had rather different ideas about the impacts of Adam's disobedience on humanity. These included Theodore of Mopsuestia, the Cappadocian Fathers, Athanasius of Alexandria, Methodias of Olympus, and Hilary of Poitiers, among others.

Even so, Augustine's formulation did eventually become a mainstay of the Latin church and continues to be so for the modern western church. It thus has the force of tradition and, as with all long-held traditions, should be deeply respected. At the same time, Christian doctrine is not immutable, whatever traditionalists may say. It can and does change. Sincere Christian theologians can disagree on what correct doctrine is, and they may change their views over time. One has only to look at the more than 30 major church schisms that have occurred since the first century and the proliferation of Protestant denominations since the Reformation in the 16th Century. All of these branches of Christianity grew at least to some degree out of disagreements over doctrine.

If there were no conceivable way to reconcile biological evolution with Christianity without eviscerating the core doctrine of Christ, then the strong reactionary views of traditionalists would be warranted. But in fact there is a way forward, an approach that maintains the integrity of the doctrine of Christ and which is rooted in the views of the earliest theologians of the impact that Adam and Eve's disobedience had on all humanity.

Views of original sin prior to Augustine

The first available reference to the story of Adam's sin as it relates to the spiritual condition of all humans is found in the writings of

Justin Martyr in the early 2nd Century. He saw Adam's sin as nothing more than the prototype for sins committed by all others. There was no notion of some errant nature passed from Adam to his offspring, and he expressed the view that every person is responsible only for the sins that he himself commits, not Adam's sin. Similar views were held by Tatian and Theophilus of Antioch.

With Irenaeus of Lyons in the latter part of the 2nd Century we see the first suggestion of some form of connection between all humans and Adam. He envisioned an element of solidarity or identity of the whole human race with Adam, though he did not explain how it occurred or precisely what it meant. In the 4th Century, Ambrosiaster was more explicit in his belief about such solidarity, while the writings of Athanasius of Alexandria are less precise, presupposing but not explicitly stating such a connection.

Several early theologians went beyond the simple fact of our connection to Adam to address the issue of our participation in Adam's guilt. However, most maintained that we are not guilty of Adam's sin, contrary to Augustine. Both Athanasius of Alexandria and Ambrosiaster took this position, as did Clement of Alexandria, Cyril of Alexandria, Tertullian, Ambrose of Milan, Theodore of Mopsuestia, and Theodoret of Cyrrhus. Didymus the Blind is one of the few whose writings indicate a belief that all humans are guilty of Adam's sin. John Chrysostom did not explicitly say that all humans are guilty of Adam's sin, but he did say that we each incur a debt due to Adam's sin.

There was widespread belief in the transmission of some errant nature from Adam to his offspring among theologians before Augustine. However, the terms used to describe this inheritance varied considerably, with subtle differences of meaning. Irenaeus and Cyprian of Carthage said that we inherit Adam's sinfulness, while Tertullian, Ambrosiaster, Ambrose of Milan, and Cyril of Alexandria described it as the inheritance of a corrupted nature. Others described the inheritance in more biological or psychological terms, such as a disease (Gregory of Nyssa), a disordered sensuality (Clement of Alexandria), or a bias towards sensuality (Methodias of Olympus). Notably, while some described the inheritance as an inclination to sin (for example, Theodore of Mopsuestia), there is almost no concept of a thoroughly irresistible inclination to sin among theologians before Augustine. Quite to the contrary, they typically emphasized our free will to resist the sinful inclinations inherited from Adam, though most also conceded that we are prone to give in to those inclinations and that all are sinners.

The Augustinian formulation of original sin

Augustine was the bishop of Hippo in North Africa in the early 5th Century when he developed and published his views of the Fall and its implications for humanity. The Augustinian formulation of original sin has, at its simplest, three primary components:

1. All human beings are guilty of not only their own sins, but also Adam's sin.
2. All human beings have inherited from their parents, and thus ultimately from Adam, a sinful nature that inclines them to sin.
3. The sinful nature that we inherit is thoroughly irresistible.

Taken together, these three components mean that, without exception, every human being is deserving of eternal punishment in hell from the moment of conception, and no human being can please God nor do any good whatsoever without the special grace made available by Christ and obtained through faith. Augustine's formulation of original sin was consistent with the majority opinion of prior theologians on the concept of some form of inherited sinful nature, but it took the minority position that all humans are guilty of Adam's sin.

With regard to his belief that man's free will was so utterly damaged by Adam's sin that he is utterly incapable of choosing not to sin, Augustine was nearly unique among his predecessors. Free will, according to Augustine, manifests differently in four different stages of humanity's existence:

- For Adam prior to the Fall, free will meant both able to sin and able not to sin

- For Adam after the Fall, and for all his unsaved progeny, free will means unable not to sin. Man's will is free only in the sense that he can choose between sinful alternatives.
- For the saved while they remain on earth, free will once again means both able to sin and able not to sin, essentially a return to the pre-Fall state.
- For the saved in heaven, free will means unable to sin. Glorified man's will is free only in the sense that he can choose between sinless alternatives.

The will of every human being prior to their salvation is immovably bent towards sinfulness, and thus without exception every human being is in need of the special saving grace afforded only through faith in Jesus Christ in order to gain the ability to not sin and to please God. Without this special grace, humans can do nothing but sin.

Augustine advocated other concepts that did not become official church doctrine, either because there was no need (they helped support the three primary components), or because insufficient agreement about them could be garnered among other theologians. For instance, Augustine promoted the concept of double predestination under which God chooses not only those who will be saved but also those who will be consigned to an eternity in hell. It was a logical extension of his views of free will: those who are saved are those to whom God gives the ability and the desire to place their faith in Christ. We play no part in God's decision about whom He extends this grace to. Likewise God chooses to whom He will not extend this grace. Augustine believed that these divine decisions were not merely based on foreknowledge of what each person will choose, but rather determined what each person will choose.

Augustine also placed great emphasis on the role of sexuality in our sinfulness. He believed that the carnal excitement of sex was the means through which sin was transmitted from parents to children. He also believed that the sinful inclination that we inherit from our parents is concupiscence, a fervent desire or longing that most often manifests as sexual desire. Unsurprisingly, Augustine scorned sex generally, seeing in it the worst of our inherited sinfulness. In fact, he went so far as to say that sex for the purpose of pleasure rather than for the purpose of procreation is sin.

Augustine's formulation of original sin was more detailed and reasoned out than that of his predecessors. This is unsurprising as the particulars of original sin were of little interest to most theologians until the 4th Century. Their focus was instead on characterizing and defending the threefold nature of the Trinity, the twofold nature of the Christ, the nature of the Atonement, baptism, the Eucharist, and the organization and authority structure of the church. When they did address original sin (though the term itself was Augustine's), their views were rather heterogeneous.

Responses to Augustine's formulation of original sin

The Augustinian formulation of original sin was first accepted as the correct view of the spiritual condition of humanity at the Council of Carthage in 418 CE. The council had been called to formally address the controversy over Pelagianism. It was in fact this controversy that had compelled Augustine to precisely define and expound upon his formulation of original sin over the course of the previous decade or so. Prior to this controversy, Augustine's writings had alluded to some of the elements of what he would later call original sin, but had provided few details.

Pelagius had been advocating a decidedly different view of human nature under which we are not guilty of Adam's sin, we are capable of exercising our free will to not sin, and we are able to achieve moral perfection through our own efforts without any special help from God. Pelagius was concerned about what he perceived as laxity among Christians who, under the umbrella of original sin, saw themselves as having no capacity for self-improvement and therefore no responsibility to make any effort. He also believed that too many Christians were relying on Christ's forgiveness without marrying it to the pursuit of right conduct, potentially threatening the authenticity of that forgiveness. Augustine in turn saw Pelagianism as antithetical to the gospel of grace and as rendering Christ unnecessary. The over 200 bishops in attendance at the Council of Carthage agreed, and developed several canons directly anathematizing the teachings of Pelagius and indirectly affirming Augustine's views.

The canons of the Council of Carthage did not eliminate debate over Augustine's formulation of original sin, however. It continued for more than another century, prompting another council held at Orange in 529 CE. And while the Latin church viewed this second council as concluding the matter, views of original sin did not thereby fixate precisely upon Augustine's formulation for all Christians everywhere. Medieval theologians such as Anselm of Canterbury and Thomas Aquinas continued to speculate on the nature of original sin, and the Roman Catholic Church felt compelled to revisit the topic at the Council of Trent in 1546 CE in the context of its response to the emerging Protestant movement.

While there were novel elements to Augustine's rationale for original sin, he maintained that the core doctrine was nothing new. As he said in *On Marriage and Concupiscence Book II*,

> *I am not the one who has invented original sin which the Catholic faith has believed from antiquity. (per Quasten)*

But in the immediate wake of the Council of Carthage, some theologians continued to voice opposition to Augustine's formulation of original sin. For instance, Vincent of Lerins said that Augustine's formulation of original sin represented a deviation from traditional church teaching. In his *Commonitorium* written in 434 CE, he laid out the criteria that all legitimate church doctrines must meet:

> *We must hold that which has been held everywhere, always, and by all, that is, we must be guided by Universality, Antiquity, and Consent.*

These criteria, he believed, were not met by Augustine's formulation of original sin.

Such views were also held by many bishops. At the conclusion of the Council of Carthage in 418 CE, Bishop of Rome Zosimus issued a circular letter to all the bishops under his jurisdiction ordering them to subscribe to this now official church doctrine. Eighteen refused, and were subsequently removed from their positions. According to Bethune-Baker, many other bishops likely signed the letter unwillingly.

Theodore of Mopsuestia explicitly rejected Augustine's formulation of original sin, responding to it with a book entitled *Against the Defenders of Original Sin.* He took particular issue with Augustine's position on free will, believing instead that Adam's sin did not eliminate our moral capacity not to sin. On this specific issue he wrote a treatise entitled *Against Those who Say that Men Sin by Nature and not by Will.* Mark the Hermit held similar views, believing that every sin is the result of our own free will, not of some corrupted nature inherited from Adam.

Julian of Eclanum was particularly vocal in his opposition to Augustine, calling his concept of sin absurd. He also analyzed the passages of Scripture that Augustine used, concluding that they do not justify the existence of original sin.

On a wider scale, the Greek church largely rejected Augustine's formulation. While agreeing that Adam's sin generated sinful inclinations in all his progeny, they did not see those inclinations as utterly irresistible. Neither did they believe that all humans are guilty of Adam's sin, but only their own sins. As the Council of Carthage was comprised of bishops from the provinces of Africa and Spain, it was viewed by most Greek Christians as a local meeting not necessarily applicable to nor representing the views of the entirety of the church.

Conclusion

Over the centuries, faithful Christians have relied upon traditional views of doctrines to provide the strongest connection possible to the teachings of the first apostles. New ideas are generally tested against those held by the esteemed theologians and thinkers that have gone before us, with older views typically taking precedence over more recent views. But new information - archeological or paleontological evidence, for instance - can cast traditional views in a different light and prompt their reconsideration. Regarding the doctrine of original sin, knowing the tangled history of the Augustinian formulation and the many dissenters which existed in his time and afterwards are also relevant factors. Any re-examination of this doctrine is bound to elicit skepticism at first, but faithful Christians can proceed with integrity knowing that the core doctrines of the universality of man's sinfulness and his unqualified need for a Savior remain unaltered under the Revised Origins Paradigm.

Chapter 11

The Case for and Against Original Sin

There is no place in the Bible that explicitly and unambiguously lays out a doctrine of original sin. Instead, it has been derived from an assemblage of multiple passages, combined with inferences and rationale outgrowth arguments. This is one reason it was not fully developed until the 5th Century, and why there was and still is debate over its particulars.

Of all the relevant Bible passages, the most comprehensive and oft-cited is that found in Romans Chapter 5:

> *12 Therefore, just as sin entered the world through one man, and death through sin, and in this way death came to all people, because all sinned—*
>
> *13 To be sure, sin was in the world before the law was given, but sin is not charged against anyone's account where there is no law.*
>
> *14 Nevertheless, death reigned from the time of Adam to the time of Moses, even over those who did not sin by breaking a command, as did Adam, who is a pattern of the one to come.*
>
> *15 But the gift is not like the trespass. For if the many died by the trespass of the one man, how much more did God's grace and the gift that came by the grace of the one man, Jesus Christ, overflow to the many!*
>
> *16 Nor can the gift of God be compared with the result of one man's sin: The judgment followed one sin and brought condemnation, but the gift followed many trespasses and brought justification.*

17 For if, by the trespass of the one man, death reigned through that one man, how much more will those who receive God's abundant provision of grace and of the gift of righteousness reign in life through the one man, Jesus Christ!

18 Consequently, just as one trespass resulted in condemnation for all people, so also one righteous act resulted in justification and life for all people.

19 For just as through the disobedience of the one man the many were made sinners, so also through the obedience of the one man the many will be made righteous.

20 The law was brought in so that the trespass might increase. But where sin increased, grace increased all the more,

21 so that, just as sin reigned in death, so also grace might reign through righteousness to bring eternal life through Jesus Christ our Lord.

In general terms this passage describes some sort of connection between humanity and Adam, and between our spiritual state and Adam's disobedience. This connection is repeated by Paul in his first letter to the Corinthians.

For since death came through a man, the resurrection of the dead comes also through a man. For as in Adam all die, so in Christ all will be made alive. (1 Corinthians 15:21-22)

These passages are treated piecemeal below as they relate to each of the three components of Augustine's formulation of original sin.

While there are a variety of issues associated with deriving a doctrine of original sin from the Bible, what is not in question is the general sinfulness of humanity. There was essentially no disagreement among theologians in the first five centuries about this, and it is just as true under the Revised Origins Paradigm as it is under the Traditional Origins Paradigm. Bible passages affirming this truth abound.

Do not bring your servant into judgment, for no one living is righteous before you. (Psalm 143:2)

> *The hearts of people, moreover, are full of evil and there is madness in their hearts while they live, and afterward they join the dead. (Ecclesiastes 9:3b)*

> *for all have sinned and fall short of the glory of God (Romans 3:23)*

> *As for you, you were dead in your transgressions and sins (Ephesians 2:1)*

> *If we claim to be without sin, we deceive ourselves and the truth is not in us. (1 John 1:8)*

Other passages trace the general sinfulness of humanity to some inherent feature of our souls that drives us to sin.

> *For it is from within, out of a person's heart, that evil thoughts come—sexual immorality, theft, murder, adultery, greed, malice, deceit, lewdness, envy, slander, arrogance and folly. (Mark 7:21-22)*

> *For I know that good itself does not dwell in me, that is, in my sinful nature. For I have the desire to do what is good, but I cannot carry it out. For I do not do the good I want to do, but the evil I do not want to do—this I keep on doing. Now if I do what I do not want to do, it is no longer I who do it, but it is sin living in me that does it. (Romans 7:18-20)*

> *This is especially true of those who follow the corrupt desire of the flesh and despise authority. (2 Peter 2:10a)*

The doctrine of original sin takes as a given that we are sinners and that we have some underlying condition which drives us to sin, and then seeks to explain its origin and expound upon its implications for our need for a Redeemer.

Adam's guilt is our guilt

Under the Augustinian formulation of original sin, every person without exception is guilty of not only his own sins, but also Adam's sin. By definition, then, every person is worthy of eternal damnation from the moment of conception and is wholly dependent upon the saving grace offered through Christ.

There are several arguments supporting this position. The first is the mere observation that everyone dies, and since the wages of sin is death (Romans 6:23a), ipso facto every person must be guilty of sin. In the case of infants or children prior to the age of reason who cannot have committed any personal sins, there must be some other source for the sinfulness that causes their death. That they must be guilty of Adam's sin is the logical conclusion.

This argument relies on the premise that Adam was created immortal and that God actively imposed death as the punishment for his sin. And as it was for Adam, so it is for all his progeny: we die because we are guilty of sin, and if it were not for that guilt, we would not die. But as discussed in Chapter 5, the Revised Origins Paradigm asserts that Adam was born mortal and that death was his natural end, to be prevented only by the power afforded by the fruit from the tree of life. God's punishment for Adam's sin did not involve actively imposing death, but rather removing the means for preventing death. If physical death is the natural end of every human being, then the fact of physical death cannot be prima facie evidence that we are all guilty of Adam's sin. It remains true that the punishment of death laid on Adam is laid on us all as discussed further below, but we cannot conclude from this fact that we are all guilty of Adam's sin.

The second argument supporting the position that we are guilty of Adam's sin is derived from the following passage:

> *Nor can the gift of God be compared with the result of one man's sin: The judgment followed one sin and brought condemnation, but the gift followed many trespasses and brought justification. ... Consequently, just as one trespass resulted in condemnation for all people, so also one righteous*

act resulted in justification and life for all people. (Romans 5:16,18)

In the wider passage of Romans 5:12-21 as well as in 1 Corinthians 15:12-22, Paul more commonly describes the impact of Adam's sin on all humanity as death. However, in this passage Paul uses the word condemnation rather than death. The underlying Greek word *katakrima* means a damnatory sentence, suggesting that it applies to a whole person rather than only to one's physical body. Additionally, condemnation is contrasted with justification in the passage above, indicating that it represents separation from God rather than merely physical death.

However, this passage says nothing about Adam's guilt or its extension to all of humanity. Rather, guilt is an inference based on the supposition that God in his justice would never condemn anyone who is not guilty of sin. But in fact it is quite possible that condemnation can be applied to those who are not guilty. As described in Chapter 7, Adam was intended by God to be the means through which He would complete His work to bring order to the earth and draw all people to Himself. Adam carried an enormous responsibility, and his choices would affect not just himself but all others as well. Thus in disobeying God, Adam interrupted God's intended process and, as the representative of all humanity, the punishment that fell on him fell also on everyone else despite the fact that only Adam sinned. This is the meaning of verses 15 and 17 of Romans Chapter 5 as well as in 1 Corinthians 15:12-22 where Paul makes plain that Adam's punishment is transmitted to all of humanity without saying anything about the transmission of his guilt.

It may also be the case that condemnation refers to Adam and Eve's exclusion from the garden of Eden and the abundant life with God that the garden represented. When this occurred, all other humans on the earth at that time would also have been excluded from the garden. They were condemned because of Adam's sin in the sense that God's intended means for connecting all people to Himself was lost.

Finally, there may be an intermediate step not explicitly mentioned in this passage. Rather than inferring that all of humanity is guilty of Adam's sin because all are condemned, it may be the case that all are condemned because all sin - verse 12 makes this explicit - and all sin because Adam set in motion a pattern of disobedience that all followed. We are condemned because we are guilty of our own sin, not because we are guilty of Adam's sin, but even our own sin can be traced back to Adam because his was the first or "original" sin. In this way, Adam's sin resulted in condemnation for all without it being necessary to conclude that all are guilty of Adam's sin.

There are two additional arguments supporting the position that we are guilty of Adam's sin. These are the concept of seminal presence and the tradition of infant baptism in the early church. These topics have wider implications and so are treated separately below.

The chief weakness in the concept that all of humanity is guilty of Adam's sin is that it runs counter to Biblical passages which make it clear that every person is responsible only for his own actions. For instance,

> *The one who sins is the one who will die. The child will not share the guilt of the parent, nor will the parent share the guilt of the child. The righteousness of the righteous will be credited to them, and the wickedness of the wicked will be charged against them. (Ezekiel 18:20)*
>
> *God will repay each person according to what they have done. (Romans 2:6)*
>
> *The sea gave up the dead that were in it, and death and Hades gave up the dead that were in them, and each person was judged according to what they had done. (Revelation 20:13)*

Several early theologians made this very argument in response to the idea that Adam's guilt is our guilt.

Supporters of the view that every person is guilty of Adam's sin rarely mention one rather awkward implication: if the guilt of Adam can be imputed to his offspring, there is no reason that the same cannot be true of all men. Each person must be guilty of the sins of his parents, his grandparents, his great-grandparents, all the way back to Adam. If it is difficult to reconcile the verses above with the attribution of Adam's guilt to all people, it is absurdly difficult to do so if every individual is guilty of all sins from all his ancestors.

Seminal presence

As discussed in Chapter 10, several early theologians envisioned some form of identity or solidarity between all of humanity and Adam, though most before Augustine did not specify what this meant. It was Ambrose of Milan in the late 4th Century, Augustine's teacher, who made the connection more precise by originating the concept of seminal presence or seminal identity of every individual with Adam. In this concept, since every human being is the product of sperm from his father (there was no awareness at that time of the contribution of egg by the mother), we were all present in our father's body prior to our conception. And our father was likewise present in our grandfather's body at one time. The procreative chain leads all the way back to Adam, such that every person can be said to have existed in the body of Adam. When Adam sinned, then, all of humanity was present with him and all participated in that sin. Having participated in Adam's sin, we are as guilty of that sin as Adam was.

The concept of seminal presence is most directly supported by an extrapolation from Paul's description of the relationship between Levi and Abraham.

> *One might even say that Levi, who collects the tenth, paid the tenth through Abraham, because when Melchizedek met Abraham, Levi was still in the body of his ancestor. (Hebrews 7:9-10)*

The idea that Levi was present in his ancestor Abraham is taken by Paul to mean that Levi acted along with Abraham in paying the tenth to Melchizedek. What Abraham did, Levi also did because Levi was in Abraham's body. As one might expect, the concept of seminal identity also played a role in some early discussions about our inherited sinfulness.

The difficulty in applying this passage from Hebrews to all of humanity lies in its introductory phrase "One might even say that". It suggests that what follows is metaphorical, not to be taken literally. The entirety of Hebrews Chapter 7 is Paul's attempt to show the preeminence of Jesus as the ultimate high priest. In this context, Paul uses the connection of Levi to Abraham in support of the more central point that Jesus did not emerge from the Levitical priesthood just as Melchizedek did not emerge from the Levitical priesthood. Paul emphasizes the metaphorical nature of this comparison when a few verses earlier he describes Melchizedek as

> *Without father or mother, without genealogy, without beginning of days or end of life, resembling the Son of God, he remains a priest forever. (Hebrews 7:3)*

These things describing Melchizedek were not true in actuality, but Paul nevertheless uses them to paint a picture of Melchizedek's role as priest that provides a parallel with Jesus' role as priest. The most internally consistent treatment of Paul's discourse would mean that his description of Levi being in the body of Abraham was also not true in actuality, but was merely used in an abstract way to support his larger point about the priesthood of Jesus.

To employ the passage from Hebrews 7:9-10 in support of the concept that all of humanity was actually, physically present with Adam when he sinned is a considerable stretch of its original intent. Moreover, it elevates what was a minor and unique point in Paul's chapter-long description of the priesthood of Jesus to a foundational principle in the concept of original sin.

It is also worth noting that, from a metaphysical point of view, seminal presence is quite problematic. To say that a human being is the same as the sperm from which he was conceived, that one's thoughts, emotions, will, and spirit - everything that makes a person a person - were present in the dab of proteins and DNA making up sperm is nonsensical. One would at least need to speculate that some essential element of personhood, the germ of one's soul or the immortal spirit, was somehow present in the sperm. More than that, in fact, as that germ of personhood would need to have been present not only in one's father's sperm, but that same germ of personhood would also need to have been present in one's grandfather's sperm and great-grandfather's sperm. Effectively, Adam's sperm would need to have contained such germs of persons for the entirety of humanity despite the fact that his sperm only represented the biological beginnings of the next, single generation. This raises a host of questions that are essentially unanswerable without pure speculation: Was a germ of personhood contained only in those specific spermatozoa that ended up fertilizing an egg, or instead was such a germ of personhood present in every single spermatozoa including the 99+% which never fertilize an egg and thus never develop into a full human? Where does the germ of personhood reside before a man's body is mature enough to produce sperm? And because in reality every person is the product not only of sperm from his father but also of an egg from his mother, is only half of his personhood present in the sperm?

Augustine did not rely exclusively on Hebrews 7:9-10 in his assertion that all of humanity was present with Adam when he sinned. He also made ample use of Romans 5:12:

> *Therefore, just as sin entered the world through one man, and death through sin, and in this way death came to all people, because all sinned*

This is the modern English translation derived from a careful analysis of the original Greek. Augustine, however, did not know Greek and instead made use of a Latin translation called the Vulgate, as did all of his Latin contemporaries. Produced between the late 4th and early 5th Century by Jerome and others, the Vulgate presented Romans 5:12 differently:

> *Therefore, just as sin entered the world through one man, in whom all have sinned, and death through sin, and in this way death came to all people*

Instead of the more accurate phrase "because all sinned," the Vulgate used the phrase "in whom all have sinned." This difference was consequential as it turned what should be a connection between everyone's death and their personal sins into a connection between everyone's death and Adam's sin. It is easy to see from the Vulgate's translation how one could conclude that all of humanity was present with Adam when he sinned, and that as a result all of Adam's descendants are guilty of his sin and have been made sinners. In reality, the verse merely says that all die because all sin. The correct rendering of this verse is found in all modern translations, including the New International Version, New Revised Standard Version, New King James Version, New American Standard Bible, New Living Translation, English Standard Version, and Christian Standard Bible.

The Vulgate's incorrect translation of Romans 5:12 was known in Augustine's time. Theodore of Mopsuestia and Julian of Eclanum, both Greek Christians, knew the correct meaning of Romans 5:12 and made it clear that it did not support any form of inclusion of all humanity in Adam. Julian in particular was severely critical of Augustine's use of the "in whom" Latin translation. However, the Vulgate remained popular among Latin theologians since its development, even being formally declared as having magisterial authority at the Council of Trent in the 16th Century. Not without dissentions, it should be noted. Just a few decades prior to the Council of Trent, the Catholic theologian Desiderius Erasmus of Rotterdam demonstrated that the phrase "in whom all have sinned" was indefensible.

To be sure, Romans 5:12 does make a connection between all of humanity and Adam. Specifically, it connects Adam's punishment for his sin to our punishment for our sin. Adam's sin was the first sin, and therefore the punishment of death was the first such punishment. We, like Adam, sin, and therefore we, like Adam, die. The rest of the Romans 5:12-21 passage expounds on the contrast between the disastrous consequences for humanity of Adam's sin and the victory that Jesus won over sin and death through his sacrifice on the cross.

Infant baptism

In justifying his formulation of the doctrine of original sin, Augustine relied heavily on the church's practice of infant baptism. In his view, infant baptism had been standard practice in the church since the beginning of Christianity, and was supported and taught by church Fathers. It was therefore sacred tradition and thus unchallengeable. Augustine then reasoned that the practice of infant baptism presupposed a condition that the sacrament was meant to address.

In the New Testament, baptism was the means through which sins were forgiven. But since an infant could not have committed any sins of its own, from whence came the sin for which the baptism was the cure? To Augustine, the conclusion was obvious: the infant must have within its soul some sinfulness inherited from its parents. In essence, Augustine reasoned from a pre-existing sacramental practice to the presence of original sin in all humans from the moment of conception.

With the advantage of more comprehensive scholarship into the history of baptism, scholars today know more about it than did Augustine. For instance, infant baptism was not standard practice in the church until late into the 4th Century. According to Ferguson, Pope Siricius (384-399) provided an indication of the infrequency of infant baptism in Rome when he said,

> *...whoever has vowed himself to the services of the church from his infancy must be baptized before the years of puberty and join the ministry of lectors.*

By stipulating that those who serve in the church must have been baptized before they reach puberty, Pope Siricius was accounting for the fact that infant baptism was not common practice. Were all infants of Christian parents baptized as a matter of course, there would have been no need to state that it was requirement for church service.

To be sure, infant baptism was practiced as early as the late 2nd Century, as testified by Tertullian. But there is little indication that it was the norm. On the contrary, the available descriptions we have of baptism in the early centuries strongly suggest that baptism was oriented towards adults. The Didache (1st - 2nd Century), a catechism which prescribes a variety of Christian practices, indicates that baptismal candidates should first receive detailed instruction in the moral life and should fast, and after being baptized should take part in the Eucharist. Similarly, the baptism rite itself indicates that the aspirant should be fully immersed in the water, something that would have been inappropriate if not downright dangerous for an infant. Pouring water over the head was only acceptable if there was insufficient water for full immersion.

A universal prerequisite for baptism was repentance, faith in the saving grace afforded through Christ's sacrifice on the cross and in His resurrection, and a commitment to the moral life described in Scripture. This is true not only in the early writings we have available, but also in the New Testament descriptions of baptism. John the Baptist called loudly and repeatedly for repentance prior to baptizing those who came to him. Paul explicitly connects baptism with faith in two different epistles.

> *having been buried with him in baptism, in which you were also raised with him through your faith in the working of God, who raised him from the dead. (Colossians 2:12)*

> *So in Christ Jesus you are all children of God through faith, for all of you who were baptized into Christ have clothed yourselves with Christ. (Galatians 3:26-27)*

Following Peter's preaching and appeal to observers at Pentecost, we are told that those who were baptized were those who listened to his preaching and accepted it.

> *Those who accepted his message were baptized (Acts 2:41a)*

But infants do not have the capacity to repent, fast, listen to and accept the gospel, express faith in God, or commit to living a morally pure life. Knowing this, Augustine said that the parents of an infant could take these actions on behalf of their infant. They would stand as proxies, declaring aloud in the baptismal ceremony that the infant places its faith in Christ Jesus and commits to following him for the rest of its life. Following baptism, the infant was called a "believer" based not on its own faith, but rather based on the faith expressed by its parents. But there is no Scriptural basis for such proxies, and indeed it stands in contrast to verses which explicitly say that every person is responsible for his or her own actions (e.g., Revelation 20:13).

Augustine also appealed directly to Scripture to support infant baptism. However, he was forced to point to the most ambiguous of passages as there are no references to baptism of infants or even of children in the New Testament. There are several references to whole households that were baptized (e.g., Acts 10:23-48, Acts 16:33, 1 Corinthians 1:16), and it is possible that such households included infants. But not only is the presence of infants in these houses speculative, it is not clear that they would have been baptized with the rest of the household. Indeed, in the story of Cornelius in Acts, his household heard and accepted the word preached by Paul and spoke in tongues, actions that could not have applied to infants.

There are other New Testament references to children which do not mention baptism, but could be inferred to include it. In Paul's letter to Titus, he says that an elder must be

> *a man whose children believe (Titus 1:6a)*

It might be reasonable to conclude that children must be baptized in order to be considered believers. However, if the baptism of infants was routine for all Christians, it would have been superfluous for Paul to create this particular prerequisite for elders. Moreover, it is not clear that children would include infants in this passage since Paul adds that the elder's children must not be

> *open to the charge of being wild and disobedient. (Titus 1:6b)*

This is clearly a description of children of a least several years of age, not infants.

Another passage that might conceivably make a connection between children and baptism is the following:

> *Peter replied, "Repent and be baptized, every one of you, in the name of Jesus Christ for the forgiveness of your sins. And you will receive the gift of the Holy Spirit. The promise is for you and your children and for all who are far off - for all whom the Lord our God will call." (Acts 2:38-39)*

The promise of forgiveness of sins and of the gift of the Holy Spirit includes those that are geographically removed from those immediately present ("all who are far off"). The most straightforward, parallel reading of "your children" would then be a reference to future generations, representing those who are temporally removed from those immediately present. Peter was indicating that the promise was not merely for the here and now, but also for the distant and future. There is no reason to think that his reference to children was directed at infants within the crowd listening to him.

Finally, Augustine's belief that the church fathers had always taught and supported infant baptism was also inaccurate. Some certainly did, as with Cyprian and Didymus the blind in the 3rd and 4th Centuries, respectively, both of whom believed that it conferred forgiveness of sin for the infant. Others supported infant baptism, but not necessarily for the purpose of the forgiveness of sins. In the 2nd Century, Justin Martyr said that infant baptism was necessary because infants are born with "wayward inclinations" but did not define this phrase as indicating an irresistible urge to sin as Augustine would more than two centuries later. Somewhat later, Asterius the Homilist endorsed infant baptism as a replacement for Jewish circumcision, i.e. enfolding the infant into the Christian community, and to protect the infant against demons and heresy, but he made no mention of anything resembling original sin. The author of the Apostolic Constitutions in the late 4th Century clearly supported infant baptism but made no explicit connection to forgiveness of sins. In the same timeframe, the Cappadocian Fathers went further in their support of infant baptism for the purpose of consecration to God and sanctification, all while maintaining that infants are born without sin.

More importantly, many others prior to Augustine did not believe that infants had any sin that required forgiveness, and in some cases they were explicitly against the baptism of infants for the forgiveness of sins. For instance, in the late 2nd Century Irenaeus emphasized the necessity of faith for salvation and affirmed the innocence of children by saying that those who are saved are

> *those who believed in God and continued in his love...and innocent children, who have had no sense of evil. (per Ferguson)*

At around the same time, the author of the Epistle of Barnabas argued explicitly that the souls of children are entirely innocent and that they are born without sin. In the late 2nd - early 3rd Century, Clement of Alexandria insisted that children enter the world exempt from sin and that a newborn baby has not fallen under the curse of Adam. Similar views were held by Tertullian who believed that children begin their lives as innocent, though he admitted to a bias towards sin. As described by Wiley, Tertullian's "bias" is not the equivalent of original sin as actual sin, and he did not include any element of participation in the guilt of Adam as Augustine would later do. Tertullian also emphasized that the baptism rite itself did not remove sin; only the accompanying repentance accomplished that, something that could only be done by persons of responsible age. Finally, as the time of Augustine drew near in the late 4th Century, Cyril of Jerusalem said,

> *we come sinless into this world; we sin now voluntarily. (per Wallace)*

There is one Bible passage that seems to support the idea that babies are born sinless. After king David's adultery with Bathsheba, Nathan the prophet declares that the child she bore would die as punishment. God strikes the child with illness, and although David fasts and pleads with God for the life of his son, the child eventually dies. At that point David ceases fasting and praying. In explaining his actions to his servants, David makes this statement:

> *But now that he is dead, why should I go on fasting? Can I bring him back again? I will go to him, but he will not return to me. (2 Samuel 12:23b)*

David believed that upon his death, he would go to the same place as his child. Since David saw his eternal home as being with God, he must also have thought that his child's eternal home was with God. One might reasonably conclude, then, that the child was sinless.

Augustine devised his formulation of original sin in the early years of the 5th Century at a time when infant baptism was becoming more common. This development may have been due in part to high rates of infant mortality leading to emergency baptisms for dying infants in the

2nd Century which over the following two centuries led to infant baptisms becoming more routine. While it is unclear if the purpose of routine infant baptism by the time of Augustine was always for the forgiveness of sins - as discussed above, there were other reasons to baptize infants - it is now clear that there was no universal belief in the sinfulness of infants, though some church fathers certainly believed it. Nevertheless, Augustine found in infant baptism a liturgical practice that could be brought into the service of his fight against the Pelagians concerning what role, if any, we play in our own salvation.

After Augustine's views of infant baptism and its implications for original sin had been largely accepted among Latin Christians in the early 5th Century, there remained those who found it questionable or even abominable. To those who expressed discomfort at the thought that an unbaptized infant would spend an eternity in hell, Augustine retreated to God's inscrutable wisdom and double predestination under which God chooses not only those who will be saved but also those who will be consigned to an eternity in hell. As it would turn out, the Council of Orange in 529 CE explicitly denied double predestination, undermining Augustine's use of infant baptism in his rationale for original sin. Others were outraged at the idea of infants being sent to hell. For instance, Julian of Eclanum said in a letter to Augustine:

> *Tell me then, tell me: who is this person who inflicts punishment on innocent creatures... You answer: God. God, you say! God!*

While perhaps more impassioned than many of his fellow Greek Christians, Julian nevertheless represented their general view that Augustine's formulation of original sin was unbiblical.

Adam's nature was corrupted by his sin

In addition to all of humanity being guilty of Adam's sin, Augustine asserted that every human being has inherited a sinful nature from Adam. The underlying premise for this element of his formulation of original sin is that Adam's nature was created perfect by God and was subsequently corrupted by his disobedience. Before addressing the rationale for humanity's inheritance of a sinful nature from Adam, it is important to address the rationale behind the premise that Adam's nature was corrupted by his sin.

A large part of the rationale is the result of working backwards from the conclusion that all of humanity inherits a sinful nature from Adam. How could this be unless Adam had a sinful nature? And how could Adam have a sinful nature when he was created perfect? That there must have been a transition in Adam's nature from pure and sinless to corrupted and sinful seems a logical conclusion.

However, as discussed in Chapter 6, there is little in Genesis Chapters 1-3 to suggest that Adam was created perfect, morally or otherwise. That there was a change in Adam's nature is similarly lacking any direct Biblical support. There are, however, several verses that could be interpreted as referring indirectly to a change in Adam's nature. The one that comes closest is the following:

> *And Adam was not the one deceived; it was the woman who was deceived and became a sinner. (1 Timothy 2:14)*

It is reasonable to conclude from this verse that if Eve became a sinner by eating from the tree of knowledge, she must not have been a sinner beforehand. But traditionalists take this one step further to conclude that Eve's nature had changed. By sinning once, she corrupted her nature so that she was inclined to sin again in the future. She became one who regularly sins. But this verse could alternatively be taken to mean merely that, once Eve had sinned, she now fell into a new category. She sinned by doing what God had expressly commanded against, therefore she was now a sinner. Even if she never sinned again, she would still be categorized as a sinner. It is not necessary to conclude that her single act of disobedience set her on an inescapable path of perpetual sin.

Paul is the only New Testament writer who references Adam and Eve by name, and he generally treats Adam, not Eve, as carrying the primary responsibility and thus guilt for the Fall. But in this verse he singles out Eve as having initiated the pair's disobedience. From the wider passage in 1 Timothy Chapter 2 it is clear that he focuses on Eve because he is providing directions to Timothy on the proper role of women in the church. His intention is not to provide insight into the impacts of the Fall on humanity. If that had been his intention, Paul would certainly have made Adam his focus rather than Eve in keeping with his discussions elsewhere of the impact that Adam's sin had on humanity.

Other Bible passages that might be taken as evidence that Adam's nature changed as a result of his disobedience are even more ambiguous. For instance:

> *sin entered the world through one man (Romans 5:12a)*

If Adam's sin of eating from the tree of the knowledge of good and evil was the first sin to have existed in the world, then neither he nor Eve (nor, for that matter, any other people who may have been on earth at that time) could have committed any sins prior to that event. This, in turn, might suggest that Adam and Eve were previously morally pure, with no inward inclination towards sin. When they disobeyed God, sin entered the world not only as the first sinful act, but also as the beginning of a sinful state of affairs on the earth. The moral atmosphere had changed, so to speak. Traditionalists identify the locus of that change as being within Adam himself.

The link between Romans 5:12 and Adam's transition from being pure and perfect to having a corrupted, sinful nature is not the most straightforward reading, being the result of a string of inferences. This verse could more easily be interpreted to mean only that Adam's sin was the first sin to have occurred in the world. This would be more consistent with the very next verse.

> *To be sure, sin was in the world before the law was given, but sin is not charged against anyone's account where there is no law. (Romans 5:13)*

Paul does not specify which law he is speaking of here. The natural progression from verse 12 to verse 13 would suggest that he is speaking of the law given to Adam and Eve forbidding them to eat from the tree of the knowledge of good and evil. However, verse 14 references Moses, which could mean that the law of which Paul speaks in verse 13 is the Mosaic law. Regardless, Paul is asserting that without a clear command from God which sets boundaries for his children, they are not punished for having crossed those boundaries. Adam and Eve could have sinned prior to God's command, but if so their sin would not have been worthy of punishment because they were ignorant of any boundaries that they should not cross. Jesus makes essentially the same point when referring to those who rejected His message.

> *If I had not come and spoken to them, they would not be guilty of sin; but now they have no excuse for their sin. (John 15:22)*

In effect, Adam's disobedience was the first sin of consequence in the world because God's commandment to Adam and Eve was the first law. Romans 5:12 need not be read to suggest anything more than this.

More problematic than the lack of any explicit Biblical support for the premise that Adam's nature changed as a result of his disobedience is how a perfect being could even conceive of sin let alone be capable of sinning. If our sinful nature is what creates the desire to sin, then it stands to reason that absent a sinful nature there should be no desire to sin. And yet, prior to the Fall, Adam desired to sin without having a sinful nature. Theologians throughout history have highlighted this difficulty without being able to offer any satisfactory explanation. R.C. Sproul admits as much in his book *Chosen by God.*

> *But Adam and Eve were not created fallen. They had no sin nature. They were good creatures with a free will. Yet they chose to sin. Why? I don't know. Nor have I found anyone yet who does know.*

In light of this dilemma, traditionalists typically focus on the fact that Adam and Eve had free will and thus were able to choose to sin. But this sidesteps the core question of how it is possible to choose what one cannot conceive of and has no desire for. The Revised Origins Paradigm resolves this dilemma by asserting that Adam and Eve were not perfect before the Fall, and that those imperfections led them to mistrust God's goodness and provision and to eat from the tree of knowledge.

We inherit a sinful nature from Adam

In western Christianity, descended from the Latin arm of the early church, the doctrine of original sin is commonly viewed as synonymous with the concept that every individual inherits a sinful nature from Adam. The other two elements of Augustine's formulation - that we are guilty of Adam's sin in addition to our own sin and that we are incapable of doing anything but sin without special grace from God - are often viewed as secondary and in some churches may not be acknowledged at all. The inheritance of a sinful nature therefore often takes center stage in discussions about the impact that Adam's sin had on humanity. It is also the area of greatest concern among Christians who see biological evolution as incompatible with the universal need for Christ; if not everyone is a descendant of Adam, then not everyone inherited his sinful nature. As will be discussed in Chapter 13, the Revised Origins Paradigm retains the concept that we all inherit an inclination towards sin, though not as a direct result of Adam's disobedience in the garden of Eden. In preparation for that discussion, this section presents the biblical evidence for the inheritance of a sinful nature from Adam and some of the difficulties that arise.

As for all elements of the Augustinian formulation of original sin, the classic proof texts are Romans 5:12-21 and 1 Corinthians 15:21-22. Most of what these passages describe, however, is humanity's inheritance of the punishment of death. While one might infer the inheritance of a sinful nature from explicit statements that we inherit the punishment of death, this link is weak; death is the punishment for sinful acts, not the punishment for having a sinful nature. As discussed above, Romans 5:12 makes it clear that all die because all sin. The reason that we sin is another matter.

Only one verse in these two passages has the potential to speak directly to an inheritance of a sinful nature:

> *through the disobedience of the one man the many were made sinners (Romans 5:19a)*

Paul makes a clear causal connection between Adam's sin and the sinfulness of humanity. What is not clear, however, is precisely how Adam's sin does this. Traditionalists read this verse to say that Adam's sin corrupted his nature so that he became inclined towards sin, and that this inclination was then passed to all his progeny. But it would also be reasonable to conclude that Adam's disobedience, and the underlying motivation of doubt and distrust of God, created a distorted social environment that influenced his children and grandchildren. Adam's progeny became sinners because they followed in Adam's footsteps, and were inclined to do so because Adam provided them an example to follow.

These two different posited mechanisms through which humanity became sinful was a significant point of contention between Augustine's supporters and detractors. The disagreement fell largely along Latin-Greek lines. Those who saw little or no Biblical support for the concept of the inheritance of a sinful nature were those who instead saw humanity as following Adam's poor example.

There are other Scripture passages which have been used in support of humanity's inheritance of a sinful nature from Adam. Those most directly relevant appear to say we are born sinful. The classic example is the following.

> *Surely I was sinful at birth, sinful from the time my mother conceived me. (Psalm 51:5)*

Since infants cannot have committed any sins of their own, their sinfulness must have come from their parents. It is then logical to extrapolate backwards all the way to Adam, and to conclude that Adam's sinful nature must have been transmitted to every person. However, a closer look at these passages reveals that their meaning is not unambiguous.

Psalm 51 is king David's lament over his sin with Bathsheba. It is strongly emotional, and his overall goal is to humble himself to the utmost.

> *My sacrifice, O God, is a broken spirit; a broken and contrite heart you, God, will not despise. (Psalm 51:17)*

In this context, David uses some extreme language in his confession of guilt. For instance:

Against you, you only, have I sinned (Psalm 51:4a)

This is most certainly not true, as David also sinned against Bathsheba and her husband Uriah. But by narrowing the scope of his guilt, David emphasizes his contrition before the God of the universe. David's goal was not complete accuracy, but rather complete repentance. Similarly, he describes his new awareness of his sinfulness this way:

For I know my transgressions, and my sin is always before me. (Psalm 51:3)

Note that David speaks here not just of one transgression, but multiple transgressions. It would seem that the sin of which he speaks is not merely his sin with Bathsheba, but his whole history of sin. Prompted by the realization of the gravity of his sin with Bathsheba, all his past sins are coming to mind. He feels the weight of them, and knows that he has no hope before God except to confess them all and rely on His mercy. It is in this context that he makes the statement often cited as support for humanity's inheritance of a sinful nature.

Surely I was sinful at birth, sinful from the time my mother conceived me. (Psalm 51:5)

Given the overall context, there is good reason to believe that David was using hyperbole here. A paraphrase might be "I am sinful and have always been sinful." It is the impassioned confession of someone who realizes that there is no value in hiding anything, and that in the harsh light of God's goodness there are none who are truly good.

That verse 5 is meant to be hyperbolic is further supported by verse 6 where the hyperbole is even more obvious.

Yet you desired faithfulness even in the womb; you taught me wisdom in that secret place. (Psalm 51:6)

Faithfulness and wisdom are not attributes typically associated with an unborn baby. David uses this language to describe God's intended relationship with his children from the very beginning of their lives, not to describe what is actually true of an unborn baby. A consistent treatment of Psalm 51 overall would lead one to conclude that verse 5 is likewise not meant to be taken literally.

This same type of hyperbole is found in other verses which discuss infants.

> *from infancy you have known the Holy Scriptures (2 Timothy 3:15a)*

> *from my birth I guided the widow (Job 31:18b)*

> *From the lips of children and infants you, Lord, have called forth your praise (Matthew 21:16b)*

> *Yet you brought me out of the womb; you made me trust in you, even at my mother's breast. (Psalm 22:9)*

In the face of these verses describing activities that cannot be attributed to infants in reality, one cannot assume that verses referencing the sinfulness of infants are meant to be taken literally.

This type of hyperbole is also present in other verses used in support of humanity's inheritance of a sinful nature from Adam.

> *Even from birth the wicked go astray; from the womb they are wayward, spreading lies. (Psalm 58:3)*

Infants are incapable of spreading lies, indicating that the language being used here is not meant to be taken literally. Moreover, this verse implies that a wicked adult has always been wicked, including in his teens, childhood, and infancy. This is an extreme position and unlikely to be true in many cases; an infant is not destined to become a wicked adult, but is heavily influenced by the environment in which he grows up. Nevertheless, David uses this unqualified characterization of unjust rulers to emphasize the breadth of their wickedness. Given these factors, this verse cannot be taken as prima facie evidence that all infants are born sinful.

Another verse sometimes used in support of the concept that we inherit a sinful nature from Adam is the following:

every inclination of the human heart is evil from childhood. (Genesis 8:21b)

It is not clear if the reference to childhood includes infants or instead is limited to children of responsible age. In other verses a distinction is made between children and infants, suggesting that the author of Genesis 8:21 could have explicitly referenced infants if he had intended them to be included.

He also put to the sword Nob, the town of the priests, with its men and women, its children and infants, and its cattle, donkeys and sheep. (1 Samuel 22:19)

Now this is what the LORD God Almighty, the God of Israel, says: Why bring such great disaster on yourselves by cutting off from Judah the men and women, the children and infants, and so leave yourselves without a remnant? (Jeremiah 44:7)

One might also consider whether God was referring to every single human being in Genesis 8:21, or only some humans. After all, God says this just after Noah's ark had finished its journey, and

Noah was a righteous man, blameless among the people of his time, and he walked faithfully with God. (Genesis 6:9b)

In light of this verse, it would be incongruent for God to say that Noah's heart was evil from childhood. And if Noah is an unstated exception to the general statement made in Genesis 8:21, then there is no reason not to believe that there may be other exceptions as well.

There is one additional passage that appears to point to the inheritance of a sinful nature. In Job's response to his friends' accusations that his sin was the cause of his misfortune, Job admits his general sinfulness even while maintaining his innocence of any intentional sin. While doing so, he makes this statement:

Mortals, born of woman, are of few days and full of trouble...Who can bring what is pure from the impure? No one! (Job 14:1,4)

This passage is part of Job's fourth speech, which spans Chapters 12 through 14. One of the themes of this speech is Job's observation that all of humanity is frail and helpless before the God of the universe, and Job pleads with God to leave him alone. He describes his position before God this way:

> *Will you torment a windblown leaf? Will you chase after dry chaff? (Job 13:25)*

In Chapter 14 Job continues this theme by describing the life of man as being short and difficult, and not worthy of God's attention. It is in this context that Job says that man has been born "impure," or in some translations, "unclean." In addition to being physically helpless before God, Job decries the extraordinary suffering that God has imposed on him for a condition that is common to all people.

The underlying Hebrew word for purity that Job uses can refer to physical purity, ceremonial purity, or moral purity. While the context suggests that Job means to indicate moral purity, he could have used the word righteousness instead of purity to eliminate any ambiguity. Regardless, it is not clear if Job meant to say that all of humanity is sinful from birth. He may instead be pointing to the difference between God's ultimate purity and that of creation: in light of God's purity, all of creation is comparatively impure. This seems to be what Eliphaz the Temanite is saying in Chapter 4 in the first speech given by any of Job's friends.

> *Can a mortal be more righteous than God? Can even a strong man be more pure than his Maker? If God places no trust in his servants, if he charges his angels with error, how much more those who live in houses of clay, whose foundations are in the dust, who are crushed more readily than a moth! (Job 4:17-19)*

Since there is evidence that the book of Job was the first book of the Bible to be written, it is unlikely that Eliphaz knew of the story of the rebellion of Satan and his banishment from heaven. As a result, when Eliphaz says that God "charges his angels with error" he is more likely indicating that God's scrutiny identifies faults even with His holy angels. No creature can match the purity of God no matter how pure it may be. Thus when Job says that no one can be born pure from a mother who is impure, he may merely be following Eliphaz's point that even the purest of humans is impure in the sight of God. It is not necessary to conclude from Job's statement that a sinful nature is passed from mother to child.

Other verses highlighting the general sinfulness of humanity are often seen as encompassing people of all ages, including infants. They are then used indirectly to support the concept that every person is born sinful, and consequently must have inherited that sinfulness from their parents. Several such verses are provided at the beginning of this chapter, exemplified by the following:

> *As it is written: There is no one righteous, not even one (Romans 3:10)*

Although one cannot rule out the possibility that such verses apply to infants, this interpretation ignores the fact that absolutes in Scripture such as always, never, all, everybody, nobody, and no one are often qualified and/or have exceptions, or can be deduced to be so. For instance, in the book of Acts the absolutes "all" and "no one" are used to describe the communal attitude of the first Christians.

> *All the believers were one in heart and mind. No one claimed that any of their possessions was their own, but they shared everything they had. (Acts 4:32)*

However, Chapter 5 tells the story of Ananias and Sapphira who were exceptions to the absolute statement in Chapter 4.

> *Now a man named Ananias, together with his wife Sapphira, also sold a piece of property. With his wife's full knowledge he kept back part of the money for himself, but brought the rest and put it at the apostles' feet. (Acts 5:1-2)*

> *In a similar way, passages using absolutes to say that all people are sinful cannot be interpreted with certainty to include infants.*

In the specific case of Romans 3:10, the addition of "not even one" would seem to eliminate any ambiguity as to its universal applicability. But even in this case there is reason to think otherwise. Paul is quoting Psalm 14 (with identical language in Psalm 53).

> *All have turned away, all have become corrupt; there is no one who does good, not even one. (Psalm 14:3)*

The full text of Psalms Chapter 14 paints a different picture than does verse 3 in isolation. King David begins the Psalm by directing his diatribe to fools.

> *The fool says in his heart, "There is no God." They are corrupt, their deeds are vile; there is no one who does good. (Psalm 14:1)*

The "no one" in this verse refers not to all of humanity, but to a subcategory of humanity, that of fools. Later in the Psalm, David changes to using the term "evildoers" and makes a clear distinction between them and the righteous.

> *Do all these evildoers know nothing? They devour my people as though eating bread; they never call on the LORD. But there they are, overwhelmed with dread, for God is present in the company of the righteous. (Psalm 14:4-5)*

In context, then, verse 3 that Paul quotes is not speaking about the sinfulness of all of humanity, but rather of the sinfulness of fools/evildoers. This fact is made clearer when we consider that the overall tone of Psalm 14 is David speaking against others and not including himself as being among the fools/evildoers. While verse 2 suggests that there are none "who seek God," we know that David himself did seek God, and in David's time we know of others who also sought God such as the prophet Nathan. Thus, it is not a forgone conclusion that, by quoting Psalm 14:3, Paul means to apply it more expansively to all of humanity including infants.

Original sin in toddlers

One of the tangential arguments that Augustine used, and which many have picked up on since his time, is the observation that small children exhibit behavior that might be considered the epitome of sinfulness: self-centeredness, a refusal to obey, and demanding their own way. Parents of toddlers speak of the "terrible twos" when tantrums are common and rebellion seems to color all of their interactions with the adults who have authority over them. Because toddlers lack the self-control of those of responsible age and they have not yet conformed to family and societal norms that place at least some restrictions on their behavior, that behavior is viewed as representing humanity's most core attributes. The conclusion that many have reached is that these core attributes are sinful, and their presence confirms the inheritance of original sin.

The underlying premise of this position is that, absent original sin, a toddler would be fully obedient to its parents and concerned primarily with the needs of others rather than self. As Adam and Eve did not have any children before the Fall, depriving us of any Biblical examples of sinless toddlers, this premise is based on the behavior and attitude of adults who have conformed to Biblical ideals of maturity. That is, we believe that rebellion in toddlers is in some sense unnatural, and that they would be capable of the mature behavior exhibited by God-fearing adults if only they had not inherited a sinful nature. Given what we now know about human development, this conclusion is unwarranted.

In the animal kingdom, the closer an animal species is to humans in terms of its cognitive abilities, the less a newborn offspring of that species can fend for itself and the more parenting it requires to mold it into a functioning adult. In the lower animals, newborns need relatively little rearing as they are born with all the instincts they need to find food and shelter, evade enemies, reproduce, etc. Parenting may be entirely absent. The offspring of higher animals are born (or hatch) with fewer such instincts and a greater need to learn from the examples of their parents. In the higher primates such as gorillas, chimpanzees, and bonobos, the primary instinct of newborns is suckling and crying to indicate discomfort. Their emotional and relational development is almost entirely a function of how they are raised by their parents. They are essentially a blank slate, programmed with a minimum of basic survival skills and so dependent on a nurturing environment for their development into functioning adults.

The same is true with humans, and the innate survival skills that we are born with exist to ensure that our most basic needs are met when we are the most helpless. As infants we are all about self because the alternative - being other-focused - would result in our demise. If we did not cry as a baby when we needed to be fed, we may not be fed. If we did not cry when we needed to burp, or had a rash, or were too hot, or needed the comfort of physical touch, we may not have gotten what we needed. Our selfishness is critical to our survival.

Moreover, babies are not aware of their environment as being separate from themselves. They do not regard themselves as being individuals placed into a wider world full of other individuals, each of whom have their own needs. To a baby, the universe is itself. It is selfish not because it turns away from others towards itself, but because it cannot conceive of anything but itself.

Few point to the selfishness of babies as evidence of original sin. We understand that babies must be self-oriented in order for their needs to be met. But for some reason we have come to think of toddlers as small adults who should have outgrown the selfish attitudes of their infancy. In reality, toddlers are in a transitionary stage, retaining some of the self-centered instincts with which they were born while beginning to recognize that they are individuals among other individuals. This transition takes time, and it is unreasonable to expect a toddler to have broken with the self-centeredness with which it was born, and which was critical to its survival, in some instantaneous moment prior to toddlerdom.

Toddlers are disordered jumbles of innate desires and emerging personalities, slowly coming to the awareness of themselves as distinct individuals separate from, but in relationship with, those around them. This period is strongly experimental. They are driven to investigate every possible activity, and from their blank-slate perspective all options are equally valid. Their attitudes and understanding undergo a molding and ordering process as their parents place boundaries on their behavior. In addition, the all-encompassing view of the self as all that matters in life, which was their natural state in infancy, must give way to some accommodation to the wants and needs of others. This process requires a tearing away from the self-centeredness that served them well as babies.

From the outside this tearing away appears as rebellion, full of tears and fits of outrage at every instance of denial or redirection. But in the toddler's mind he is merely struggling to make a fundamental paradigm shift. He is transitioning from who God made him to be as a baby to who God wants him to be as an adult and as a member of humanity. This process is not only necessary, but natural and to be expected. It is, in fact, unavoidable. Even under the Traditional Origins Paradigm, if Adam and Even had not disobeyed God and there were no such thing as inherited original sin, we should not expect a "sinless" toddler to behave any differently than it does today.

Moreover, toddlers cannot rebel in the proper sense of the word because they are not yet capable of understanding that their parents have authority over them. In order to rebel, one must first understand that another has authority over you and that you have an obligation to submit to that authority. What appears to be rebellion in a toddler is instead a retreat to what it has known from the beginning of its existence: needs get met by being selfish. As a child emerges from the toddler years, it slowly comes to understand not only that its parents have authority over it, but that life involves regular correction and redirection from its parents. The world is not an endless stream of options but includes boundaries. As he grows older still he learns that those boundaries are meant to keep him safe and to ensure that he grows up to become a mature adult who can function well in relationship with others.

While these revelations from the study of childhood development allow us to see toddler behavior in a different light, they do not change the difficulty that parents face as they struggle to reign in the intractable behavior of their toddlers. For Christians with a belief that suffering of every kind is the result of the fact that we live in a fallen world, it is easy to see the stress associated with parenting toddlers as being in some sense wrong. A truly perfect creation, a Christian parent might reason, would involve no such stress and raising children would be a breeze. But as discussed in Chapter 4, suffering is not incompatible with God's good creation prior to the Fall. The apparent rebellion of toddlers and the stresses that it causes for both them and their parents is in reality the expected course in the natural development of their personhood.

Humans are utterly depraved

The third component of Augustine's formulation of original sin asserts that all human beings without exception are utterly depraved, incapable of doing any good whatsoever. This view has the advantage of emphasizing our utter dependance upon God not only for our salvation but also for the special grace to live righteous lives. But it requires a certain supposition about what happened at the Fall. Rather than recognizing that all humans were created in the image and likeness of God and retain at least some elements of that likeness even if they are unsaved (as implied by such verses as Genesis 9:6), it asserts that that likeness was obliterated when Adam disobeyed God. Aside from having an immortal spirit, we are nothing like God. We have wicked hearts that are inclined towards evil at every moment.

This view of human nature carries with it an obligation to view every good act on the part of the unsaved as good in appearance only; the motivation behind a "good" act remains evil. When an atheist volunteers at a soup kitchen, it is not because he cares about those in need but rather because he wants to feel good about himself. Comforting the grieving, visiting someone in the hospital, driving a friend to an employment interview are all driven by egoism rather than by any true concern for the other.

This viewpoint is thoroughly presumptuous given that no one can perceive the heart motivations of another. It may even be elitist, driving the Christian to see himself as being morally superior to the non-Christian even in the face of palpable evidence to the contrary. Moreover, as a doctrine the total depravity of humanity is not necessary to justify our need for Christ. The smallest sin is enough to keep us out of the presence of the holy God, and thus even the most righteous person is dependent upon the saving grace offered through the shed blood of Jesus.

There are several Scripture passages that one might use to justify a doctrine of total depravity. The one that seems clearest and most authoritative comes from Jesus himself.

> *apart from me you can do nothing (John 15:5b)*

Jesus says this in the middle of a long speech to his disciples a few hours before his arrest. In the context of preparing them to continue His work after He dies, He emphasizes that their vocation is to "bear fruit." The full chapter does not appear to be speaking of any and all good works that any person might do, but rather those good works that mark a person as a disciple of Jesus and thereby provide an example to non-believers of what sort of God the disciple serves. They are God-inspired good works which require the enabling power of His Spirit to accomplish. Jesus makes clear this connection between bearing fruit and being an example for him in the same passage.

> *This is to my Father's glory, that you bear much fruit, showing yourselves to be my disciples. (John 15:8)*

A few verses later Jesus gives the only specific example in Chapter 15 of what he means by bearing fruit.

> *My command is this: Love each other as I have loved you. Greater love has no one than this: to lay down one's life for one's friends. (John 15:12-13)*

This sort of love is extreme and is only possible for those who abide in Jesus. Thus when Jesus says that "apart from me you can do nothing" it seems likely that he is not speaking broadly about all righteous acts, but more narrowly of the sort of fruit-bearing that requires God's special power and as a result is so astonishing that it catches the attention of non-believers and draws them to Jesus.

Another verse sometimes used to support a doctrine of total depravity is in the prelude to the story of Noah.

> *The LORD saw how great the wickedness of the human race had become on the earth, and that every inclination of the thoughts of the human heart was only evil all the time. (Genesis 6:5)*

At the end of the story of Noah and the ark, God repeats the same observation about the spiritual state of humanity.

> *Never again will I curse the ground because of humans, even though every inclination of the human heart is evil from childhood. (Genesis 8:21b)*

These verses would seem to leave no room for any righteous acts on the part of any human being. Yet as pointed out above, the story makes clear that Noah was an exception to these seemingly universal statements. Moreover, the Bible makes clear that righteous people have existed throughout history.

> *In this meaningless life of mine I have seen both of these: the righteous perishing in their righteousness, and the wicked living long in their wickedness. (Ecclesiastes 7:15)*

> *For the LORD watches over the way of the righteous, but the way of the wicked leads to destruction. (Psalm 1:6)*

> *The righteous perish, and no one takes it to heart; the devout are taken away, and no one understands that the righteous are taken away to be spared from evil. (Isaiah 57:1)*

There are also multiple references to specific righteous persons.

> *Then the LORD said to Satan, "Have you considered my servant Job? There is no one on earth like him; he is blameless and upright, a man who fears God and shuns evil." (Job 1:8)*

> *In the time of Herod king of Judea there was a priest named Zechariah, who belonged to the priestly division of Abijah; his wife Elizabeth was also a descendant of Aaron. Both of them were righteous in the sight of God, observing all the Lord's commands and decrees blamelessly. (Luke 1:5-6)*

Given the abundance of examples of righteous persons in Scripture, God's declaration that "every inclination of the human heart is evil from childhood" can be seen as providing a broad description of humanity that is nevertheless not applicable to every single person.

Finally, the doctrine of utter depravity conflicts with God's expectations that we are capable of resisting sin and keeping his commands. The earliest statement to this effect comes just after the Fall. Cain was angry that God had not looked with favor on his offering, and in response God tells Cain that he simply needs to do what is right. God then adds:

> *sin is crouching at your door; it desires to have you, but you must rule over it. (Genesis 4:7b)*

Surely God would not have said that Cain must rule over sin if Cain was not capable of doing so.

Later, after God had given the law to the Israelites, he pronounces a series of blessings for obedience and curses for disobedience. He then makes it clear that they are fully capable of obedience.

> *The LORD will again delight in you and make you prosperous, just as he delighted in your ancestors, if you obey the LORD your God and keep his commands and decrees that are written in this Book of the Law and turn to the LORD your God with all your heart and with all your soul. Now what I am commanding you today is not too difficult for you or beyond your reach. (Deuteronomy 30:9b - 11)*

Just a few verses later God implores the Israelites to keep the law.

> *This day I call the heavens and the earth as witnesses against you that I have set before you life and death, blessings and curses. Now choose life, so that you and your children may live (Deuteronomy 30:19)*

If the Israelites were utterly depraved and thus incapable of doing any good, it would have been illogical and cruel for God to declare that they can.

Conclusion

All human beings are sinners, and as such all are vulnerable to the wrath of the holy and just God. Their only hope is through the redeeming work of Christ on the cross. There is no disagreement on this matter between the Revised Origins Paradigm and the Traditional Origins Paradigm. How it is that they came to be sinners, however, is a separate matter.

The doctrine of original sin as formulated by Augustine posits that Adam was created perfect and then paradoxically chose to sin, unknowingly generating a sinful nature in himself which he then passed to all his progeny. As a result, every human being is constrained to sin and can do nothing but sin. Additionally, Adam's guilt was attributed to all his progeny, compounding their guilt for their own sin. Although Augustine's formulation gained widespread support in the 5th Century and continues to be the foundation for many western Christian denominations, the Biblical basis for it is not as concrete as many Christians believe it to be. In some cases the context of a Bible passage was not taken into account, while in other cases contradicting Bible passages were ignored. Occasionally a layer of intention or meaning that originated from extra-Biblical sources was attributed to a passage.

To their credit, Augustine and his supporters wanted to protect the faith from heretical views. They made the accurate observation that human beings have a powerful and persistent inclination to sin, and they worked backwards from this fact to establish a Biblical basis for it. The resulting formulation was effective at clarifying man's utterly helpless condition before God. But now, with the advantage of more comprehensive scholarship and the knowledge of man's arrival on earth through an evolutionary process, we can conclude that the Augustinian formulation of original sin is not the best explanation of man's spiritual condition, nor is it based on the most straightforward reading of the relevant Bible passages.

It remains to be shown how human beings can be so universally sinful if there was no sinful nature generated by Adam and then passed on to all his progeny. This is the subject of the next two chapters.

Chapter 12
The Many Faces of Sin

If the Revised Origins Paradigm is going to successfully explain the universal sinfulness of humanity, it must not only address the way in which humanity as a whole was adversely affected by Adam's sin, but must also address the very nature of sin itself. Chapter 8 discussed the sin that Adam and Eve committed in the garden of Eden, and there were some allusions to what is meant by sin in Chapter 10. But the doctrine of sin is complex, far more so than most realize. A more in depth review of what the Bible means by sin will help make Chapter 13's discussion of man's innate sinfulness under the Revised Origins Paradigm more coherent.

There are over fifty words in the Bible that relate either specifically or generally to sin. Synonyms that are most familiar to Christians include:

- Transgression
- Disobedience
- Iniquity
- Rebellion
- Offense
- Wickedness
- Trespass
- Unrighteousness

These words and others all mean something slightly different, and Bible authors as well as translators endeavor to fit the appropriate word to the context. Some suggest a greater degree of wickedness and intentionality than others. Some relate to the breaching of specific laws or commands while others describe a more general failure to act appropriately. And, somewhat surprisingly, occasionally a word translated as sin is not used to describe human choices at all, but rather such things as diseases and ceremonial uncleanness.

In common usage, both within and outside of Christianity, sin is taken to mean a violation of some religious standard, moral principle, or divine law. The variety of synonyms for sin used in the Bible certainly encompass such a definition, but there is more to sin than that. All usages of sin and its variants describe some form of wrongness in the choices that humans make, something that sullies relationships between man and God and between man and man. But that wrongness does not always entail deliberate rejection of God or his laws. Sin can and often does arise from mistakes, ignorance, and incapacity. As a result, the broad spectrum of wrongness in human choices that we call sin can be viewed as a subset of the disorder that permeates our world as evidenced by such things as sickness, injury, birth defects, aging, hurricanes, droughts, and predation among animals and insects. This connection between sin and disorder will be discussed in more detail in Chapter 14.

The sin of original sin

Because the doctrine of original sin is not provided explicitly in the Bible but is instead surmised from a collection of passages, there is nothing that identifies what particular sin or sins is meant to be included in the phrase original sin. In fact, there is a case to be made that original sin does not mean any particular sin at all, but rather some defect of the soul which leads to sin. This would be closer to Augustine's formulation which is focused on the transmission of some form of corruption from parent to child which, while leading to sin, is not itself sin (though some later traditions, most notably those arising from the Protestant Reformation in the 16th Century, would identify that corruption as true sin).

Even so, it is possible to identify the first sin committed by Adam and Eve in the garden of Eden as a means for better understanding the sin of greatest relevance in the doctrine of original sin. Ironically, there is not a single word in Genesis Chapter 3 that can be translated as sin or any of its synonyms; the first appearance in the Bible of the word translated as sin arises in Chapter 4 in a dialogue between God and Cain. This absence has led theologians to deduce the first sin from the dialogue between Eve and the serpent. Traditionally that first sin has been seen as pride, defined from Genesis 3:5-6 as a desire to be like God in the extreme, in order to be one's own god and be free from the control of the Creator God. Pride, as well as the disobedience and rebellion that are also sometimes used to describe the sin of Adam and Eve, is typically seen as deliberate and thus clearly damnable. As such, pride has served the church's declaration that man is the enemy of God and wholly in need of His absolution. Some theologians have also characterized pride as the prototype of all sins in that they can all be traced back to a focus on self, with pride being the quintessential form of self-centeredness.

But as discussed in Chapter 8, there is reason to believe that the first sin was instead a mistrust of God's goodness and provision. As with pride, mistrust is a form of self-centeredness wherein we place preeminent trust in our own ability to meet our needs. But it is more passive than pride, akin to being victims of our own limitations. We generally do not make a conscious choice to mistrust God, but rather fail to exercise our God-given ability to trust in Him when something challenges our belief in His care for us. In Eve's case, she gave credence to the words of the serpent which not only contradicted God's words but also implied that God was withholding a good thing from her. Doubt in God's trustworthiness defaulted to trust in self, and she acted to do for herself what she thought God could not or would not do.

Mistrust as the first sin better fits the story of God's redemptive plan than does pride. A failure to trust in God on the part of Adam and Eve must be replaced with a new trust in God on our part, specifically in his benevolence towards us as demonstrated by the incarnation and Christ's sacrifice on the cross. The solution to the human predicament - faith - reverses the origin of the problem - mistrust.

The New Testament offers additional clues as to the sin committed by Adam and Eve, though more categorically rather than specifically. These are found in two passages, both of which were discussed previously: Romans 5:12-21 and 1 Timothy 2:14. In these passages,

Paul uses four different words when referencing Adam's sin. The most common one is the noun *hamartia* and its verb (*hamartano*) and adjective (*hamartolos*) equivalents. This Greek word is by far the most common word translated as sin in all of the New Testament, and the definition most often cited is "to miss the mark." Other definitions connote similar concepts:

- To err, be mistaken
- To miss or wander from the path of uprightness and honor
- To do wrong or go wrong
- To wander from the law of God

The overall sense of *hamartia* is that there is a right way of thinking, speaking, and behaving, but that a person has gone a different way. It carries with it essentially no indication of intentionality or degree of wickedness, and is perhaps the most general descriptor of sin. This may explain its predominance in the New Testament.

The other three words that Paul uses to describe the first sin are similar to *hamartia*, but with some nuances. *Paraptoma* is translated most frequently as trespass but also as offence, sin, transgression, and fall, and is defined as a lapse or deviation from truth and uprightness, or as a misdeed. *Parabasis* is translated as transgression, breaking a command, and sin. *Parakoe* is the most specific of the three as it is always translated as disobedience. However, these three words are used only rarely in the New Testament.

Our sinful condition

While much of the Bible is concerned with individual, discrete acts of sin, it is the sinful condition of man's heart that is of greater relevance to the role that Adam's sin played in the sinfulness of all humanity. Sinful acts proceed from sinful hearts as multiple passages affirm.

> *A good man brings good things out of the good stored up in his heart, and an evil man brings evil things out of the evil stored*

up in his heart. For the mouth speaks what the heart is full of. (Luke 6:45)

but each person is tempted when they are dragged away by their own evil desire and enticed. Then, after desire has conceived, it gives birth to sin (James 1:14-15a)

There are a variety of passages throughout the Bible that refer to individuals being sinful. Some of those in the New Testament use the Greek noun *hamartia* or its equivalents that was discussed above.

When Simon Peter saw this, he fell at Jesus' knees and said, "Go away from me, Lord; I am a sinful [hamartolos] man!" (Luke 5:8, brackets added)

You may be sure that such people are warped and sinful [hamartano]; they are self-condemned. (Titus 3:11, brackets added)

There are several passages in the Old Testament that speak of sinful persons using the Hebrew word *chata* or its derivatives. *Chata* is effectively the Hebrew equivalent of the Greek word *hamartia* and is the most common word used in the Old Testament that is translated as sin.

My son, if sinful [chatta] men entice you, do not give in to them. (Proverbs 1:10, brackets added)

Woe to the sinful [chata] nation, a people whose guilt is great, a brood of evildoers, children given to corruption! (Isaiah 1:4, brackets added)

But there are a number of words and phrases that more directly identify the sinful condition of man's heart. Depending on the translation, these phrases include sinful nature, earthly nature, corrupt nature, human nature, worldly nature, natural condition, selfish desires, earthly desires, earthly body, worldly impulses, sinful flesh, flesh, and body. All of these phrases are attempts to translate one of three Greek words: *phusis, genesis,* and *sarx*. In some cases one of these words is also paired with *hamartia* (sinful).

Phusis is most often translated as nature, and conveys the inherent characteristics of someone or something. It is used to describe God's "divine nature" in 2 Peter 1:4 and "every kind of beast" in James 3:7. Paul uses *phusis* in Ephesians to describe our sinful state before God.

> *Like the rest, we were by nature [phusis] deserving of wrath. (Ephesians 2:3b, brackets added)*

Genesis is most often translated as earth or land, and is almost never used to describe people. Instead, it is used to describe the dirt upon which we stand, a country, or the planet Earth (distinguishing it from heaven). Paul describes Adam as having been created from earth.

> *The first man was of the dust of the earth [genesis]; the second man is of heaven. (1 Corinthians 15:47, brackets added)*

By contrasting Adam and Jesus, this verse provides additional insight into Adam's nature. Although the structure of this verse might imply that Jesus (the second man) did not have an earthly (physical) body like ours, we know that in fact he did. Indeed it is considered heretical to believe otherwise. This verse is therefore comparing not the physical bodies of Adam and Jesus, but rather their personhood or essential natures. Although Adam was created in the image and likeness of God, in some sense his essential nature was earthly in a way that Jesus' was not, despite the fact that they were both composed of the same physical matter. The broader context of this verse does not say precisely what this means other than to distinguish between the "natural" of Adam and all those who descended from him and the "spiritual" of Jesus and all those who are saved.

However, Paul uses the same word *genesis* in his letter to the Colossians, and in so doing he makes an explicit connection between our sinfulness and the fact that our essential nature is earthly.

> *Put to death, therefore, whatever belongs to your earthly nature [genesis]: sexual immorality, impurity, lust, evil desires and greed, which is idolatry. (Colossians 3:5, brackets added)*

Adam's essential nature was earthly, and so is ours. Moreover, our earthly nature is the source of our sinfulness. As will be discussed in more detail in Chapter 13, this fact is an important element of the Revised Origins Paradigm, in particular with regard to what it means for humanity to be sinful.

The third word used to describe the sinful condition of man's heart is *sarx*. It is typically translated as flesh, and it appears in almost every book of the New Testament. It can refer simply to the physical bodies of humans or animals such as in John's description of the incarnation.

> *The Word became flesh [sarx] and made his dwelling among us. (John 1:14a, brackets added)*

In a similar way it is also used to refer to humans.

> *In the last days, God says, I will pour out my Spirit on all people [sarx]. (Acts 2:17a, brackets added)*

But *sarx* can also refer in a more metaphysical sense to the sensuous and/or animal nature of man. There are several examples of *sarx* being used in this way.

> *The acts of the flesh [sarx] are obvious: sexual immorality, impurity and debauchery (Galatians 5:19, brackets added)*

> *For everything in the world—the lust of the flesh [sarx], the lust of the eyes, and the pride of life—comes not from the Father but from the world. (1 John 2:16, brackets added)*

In Chapter 8 of Romans, Paul provides a moderately lengthy discussion of the sinfulness of the flesh, contrasting it with a life lived according to the Holy Spirit.

> *3 For what the law was powerless to do because it was weakened by the flesh [sarx], God did by sending his own Son in the likeness of sinful flesh [sarx] to be a sin offering. And so he condemned sin in the flesh [sarx],*

6 The mind governed by the flesh [sarx] is death, but the mind governed by the Spirit is life and peace.

7 The mind governed by the flesh [sarx] is hostile to God; it does not submit to God's law, nor can it do so.

13 For if you live according to the flesh [sarx], you will die; but if by the Spirit you put to death the misdeeds of the body, you will live.

(Romans 8:3, 6, 7, 13, brackets added)

Identifying our flesh (*sarx*) as the locus of our sinfulness complements the use of earth (*genesis*) to do the same. Notably, there is nothing in the three words that the New Testament uses to describe our sinful nature to suggest that that sinfulness is in any way foreign, something bad added to an originally pure state or something good subtracted from it. One might reasonably conclude that our sinfulness is tied to the fact that we have physical bodies created from physical matter.

Concupiscence

Concupiscence is a Latin-based word that means desire, craving, or longing. It has been used to render the Greek word *epithumia*, and it played a central role in the writings of Augustine and many others all the way through the 17th Century when they discussed original sin. However, concupiscence is now considered archaic and appears in almost no modern translations, the King James Version being the most notable exception. It is thus not typically part of the lexicon used in modern discussions of sin. Even so, the treatment of concupiscence and the underlying Greek word *epithumia* by early theologians provides additional insight into how they viewed sin.

The noun *epithumia* and its verb equivalent *epithumeo* is neutral and can be used to describe both positive and evil desires as evidenced by these two passages.

> *But I tell you that anyone who looks at a woman lustfully [epithumeo] has already committed adultery with her in his heart. (Matthew 5:28, brackets added)*

> *And he said to them, "I have eagerly desired [epithumia] to eat this Passover with you before I suffer. (Luke 22:15, brackets added)*

Despite its technically neutral meaning, concupiscence came to be used to represent *epithumia* in only those cases where it meant an excessive, inappropriate, or evil desire. Lust was a common synonym in older Bible translations, but in those cases lust referred to all forms of inordinate desire. Because lust is usually interpreted by modern readers as referring narrowly to inordinate sexual desire, modern translations use lust only in cases where the context makes it clear that sexual lust was intended. In other cases the inordinate desire intended by the word *epithumia* is translated as covetousness or simply evil desire. In fact, Augustine treated concupiscence as more or less synonymous with coveting. When Paul cites the commandment "You shall not covet" in Romans 7:7, it is the Greek word *epithumeo* that he uses, the same word that would be translated as concupiscence by later theologians.

Augustine recognized that concupiscence could represent all excesses of the physical appetites, such as sex, food, alcohol, money, and power. But he placed particular emphasis on sexual lust, so much so that he has frequently been misinterpreted as having equated concupiscence and lust. As discussed in Chapter 10, he viewed sex in rather disparaging terms, seeing in it the worst of humanity's sinfulness. Even procreation was tainted by the sexual variant of concupiscence in his view. Prior to Augustine, others had held similar views. For example, Clement of Alexandria believed that we inherit a disordered sensuality from Adam and Eve, while Methodias of Olympus believed that we inherit a bias towards sensuality and are filled with carnal yearnings and sensual cravings. After Augustine's death, Cyril of Alexandria stated that the primary sin which dominates humanity is sexual desire.

In more theological terms, Augustine saw concupiscence as man's inclination to find fulfillment in fleeting material things rather than in the eternal God. Concupiscence is not only an evil desire, but a hurtful desire as it distracts man's attention away from the ultimate source of fulfillment. To this end, Augustine and other early theologians often defined concupiscence simply as disordered desire. They contrasted it with the ordered desire that Adam and Eve had before the Fall when their reason dominated over their passions, so that all desire was under their control. After the Fall, humanity's desire became uncontrolled, involuntary, and unruly. God had intended humanity to operate above the levels of beasts through God-given reason, but they fell back into animalistic passions.

In Augustine's view, concupiscence is an outgrowth of original sin and leads to sin, but is not itself true sin. Support for this view can be found in James.

> *but each person is tempted when they are dragged away by their own evil desire [epithumia] and enticed. Then, after desire [epithumia] has conceived, it gives birth to sin; and sin, when it is full-grown, gives birth to death. (James 1:14-15, brackets added)*

But Augustine's distinction between concupiscence and original sin was nuanced. Later theologians, beginning in the 12th Century and brought to the fore of theological debate by Martin Luther in the 16th Century, would collapse the two into one, treating the disharmony and disorder labelled concupiscence as the root sin with which all human beings are burdened. For Luther, concupiscence is true sin. Ultimately this debate resulted in distinctively different views of original sin between Protestant and Catholic denominations, with Catholics maintaining Augustine's position and Protestants taking Luther's position.

Philosophical abstractions of sin

There have been a number of attempts to describe sin in more abstract terms. Some of them may appear to be extra-Biblical and to draw attention away from the critical issue of the break in the relationship between man and God. But treated properly, they each have some value in helping to understand the source of sin and its propagation.

One approach is to treat all sin as having its roots in immaturity. As described in Chapter 8, it is possible that the fruit from the tree of the knowledge of good and evil was something that God wanted Adam and Eve to have when they were mature enough. Their sin, then, amounted to eating it before they were ready. That sin interrupted the maturation process that God had intended for them, fixing them in a state of immaturity. Afterwards, they and their progeny continued to act out of this immaturity.

A related approach is to treat sin as unfulfilled potential or underachievement. God created Adam and Eve with an eye towards their future role not merely as the rulers of the garden of Eden, but as rulers of all of humanity. God had intended for Adam and Eve to grow in wisdom and faithfulness until they were ready to step into that role. By doing what was forbidden, they stepped off the path on which God had placed them, effectively surrendering their God-promised futures for near-term gratification. A similar interrupted destiny would occur later with the Israelites. After being freed from Egypt, the Israelites were on their way to the promised land when they, like Adam and Eve, failed to trust God and so were consigned to wander the wilderness until all who had escaped Egypt were dead. God had a future planned for them that they failed to achieve. We see descriptions of similar interrupted destinies in Jude and the comparable text in 2 Peter Chapter 2. In reference to those who were called to serve God but had chosen not to, Jude calls them "clouds without rain" and "trees without fruit" (Jude 12), while Peter calls them "springs without water" (2 Peter 2:17).

A more transcendent approach to describing sin is to characterize it as a cosmic power or reigning force. This approach places the focus on the aggregate of all sins and envisions them as an inexorable, systemic scourge. Sin universally affects all human beings, dragging them into its realm of control like the gravitational pull of a planet.

all, both Jews and Greeks, are under the power of sin (Romans 3:9b)

Paul also uses the language of "reign" and "dominion" when describing the power of sin, suggesting something along the lines of a tyrannical ruler.

so that, just as sin exercised dominion in death, so grace might also exercise dominion through justification leading to eternal life through Jesus Christ our Lord. (Romans 5:21)

Therefore do not let sin reign in your mortal body so that you obey its evil desires. (Romans 6:12)

For he has rescued us from the dominion of darkness and brought us into the kingdom of the Son he loves, in whom we have redemption, the forgiveness of sins. (Colossians 1:13-14)

It is possible that Paul had in mind the political oppression of his day, oppression that his audience would have been familiar with, as he sought to contrast the power of sin with the power of God.

A mildly secular approach to describing sin is to treat it as a social disease. This approach begins with the observation that every person is born into an environment already saturated with sin and the effects of sin. Every culture, institution, political regime, ideology, and economic system, not to mention relationships between individuals, has been shaped by sin. No matter what potential a child may have to do good and be good, he is compelled to navigate through this corrupted world. To do so, to fit in and to function effectively, he must play along. He finds himself imitating and perpetuating the sinful structure of society even before he is aware that he is doing so. And once he does become aware, it has already become so natural, so ingrained, that it takes a monumental effort or the intervention of God to redirect him.

In perhaps the most obvious example, sinful attitudes and behaviors are passed from parents to children. What the parents demonstrate and teach, their children perceive as normal and adopt into their own worldviews. Racist parents cultivate racist children. A misogynist father produces a misogynist son. A general disregard for law results in lawless children. Apart from the inheritance of some sinful nature in our souls, there is no doubt that the unconscious imitation of our parents plays a role in the perpetuation of sin.

Once an individual develops autonomy from his parents and enters and interacts with the world at large, he finds the options constrained by the sinful choices of those who have gone before. The school he attends, the company he works for, the bank that provides the loan for his house have all been shaped in one way or another by sin, some more overtly and profoundly than others. Systems favor some at the expense of others, reward selfish ambition, and extoll the pursuit of pleasure. Unlike Adam and Eve for whom only one tree in the garden of Eden was forbidden and most options were good, for modern man it is the opposite. There is no luxury of choosing among a plethora of sinless options, and he finds himself cooperating with his sin-tainted environment whether he wants to or not.

Sin begets sin, just as a disease propagates itself. And the consequences of sin propagate as well, sometimes independently of the sin itself. A man jailed for a crime may leave his children without any financial support. They fall into poverty, and poverty has a terrible tendency to endure across generations. This may be what is meant when the Bible says

> *he punishes the children for the sin of the parents to the third and fourth generation. (Numbers 14:18b)*

Finally, sin can be described as an element of the broader disorder that is an intrinsic characteristic of our universe. As described in Chapter 2, the natural tendency of the universe is to increase in entropy or disorder. However,

> *God is not a God of disorder but of peace (1 Corinthians 14:33a)*

In Genesis Chapter 1 we see God working to create order out of disorder by separating light from darkness, the earth from the heavens, and water from land. With the ordering done that only an omnipotent God can do, He turns the rest of the task over to Adam and Eve, telling them that they are to “subdue” the earth (Genesis 1:28, NIV). Other translations say govern or subjugate, and the Contemporary English Version says that Adam and Eve were to "bring it [the earth] under your control." But Adam and Eve’s disobedience disrupts God’s plan to bring order to the earth. Disorder begins to increase rather than decrease, in both the natural and human spheres.

In the Old Testament, the goal for humanity is *shalom* which is usually translated as peace. While *shalom* includes the concepts of personal tranquility and contentment, it is considerably broader than that. It includes health, safety, prosperity, welfare, peaceable friendships, and harmony between nations. In short, *shalom* means that everything is in good working order. It is unsurprising that sin disrupts all of these expressions of order, and might reasonably be called anti-shalom in the same way that the antichrist seeks to destroy the work of Christ.

Sin as a form of disorder is also evident in the law of Moses. If a man has an unusual bodily discharge, the process of cleansing includes a requirement that he make a sin offering.

> *On the eighth day he must take two doves or two young pigeons and come before the LORD to the entrance to the tent of meeting and give them to the priest. The priest is to sacrifice them, the one for a sin offering and the other for a burnt offering. In this way he will make atonement before the LORD for the man because of his discharge. (Leviticus 15:14-15; see parallel text for female discharge at verses 29-30)*

The bodily discharge is not identified as being the result of sin, but is nevertheless treated as sin. The priest is required to make atonement "because of his discharge." There is also a law requiring atonement for a house that has mold.

> *He shall purify the house with the bird’s blood, the fresh water, the live bird, the cedar wood, the hyssop and the scarlet yarn. Then he is to release the live bird in the open fields outside the town. In this way he will make atonement for the house, and it will be clean. (Leviticus 14:52-53)*

The word "atonement" used here is the verb *kaphar*. The equivalent noun form is *kippur* which is used in the law's requirements for Yom Kippur, the Day of Atonement when atonement is made for the sins of all the people of Israel. The same word is used to describe both cleansing a person from sin and cleansing a house of mold. Sin can thus be used to describe forms of disorder beyond that which occurs in the realm of human morality.

Everything that does not come from faith is sin

This statement from Paul in Romans 14:23 defines sin not by what it is, but by what it is not. Sin stems from any posture towards God other than full confidence in His love, provision, acceptance, and approval. This includes not just outright rebellion and unbelief, but also partial confidence in God, i.e., mistrust and doubt.

Taken at face value, defining sin as any action, word, or thought that is not motivated or accompanied by faith would also mean that a host of things we would not normally identify as sin actually are. For example, anxiety, depression, fear, despair, worthlessness, disappointment, frustration, and confusion would all fall under the umbrella of sin. Although defining sin in such an all-encompassing manner runs counter to modern views - these conditions require treatment, many would say, not condemnation - it is fitting if sin is an element of disorder, the pervasive state of the world in which things are not in proper working order. Biddle describes this disordered state of humanity in more pertinent theological terms, saying that authentic humanity is Christlikeness and that anything less is sin.

> *Sin, then, cannot be reduced to violation of rules; sin must include the failure, for whatever reason and by whatever modality, to manifest God's image. Sin is not solely the prideful effort to transcend humanity, it is also living a less than fully human life as exemplified and made possible in Jesus Christ.*

Even if there is some abnormal biochemical basis or medical malady that is contributing to a person's flawed mental or emotional state, it nevertheless still represents a state of disorder that is at odds with God's intention for a perfectly functioning humanity. It is unintentional sin, possibly even inescapable sin, but still sin.

It is difficult for the modern western church to take seriously the concept that some sins may be unintentional. This could be because the root sin, the sin of Adam and Eve and the prototype of all sins, is taken to be rebellion and disobedience to the known will of God, and such sins are necessarily intentional. Some faith traditions also take the position that all sins are intentional because every person was created in the image of God and so has an innate knowledge of His existence and His moral requirements. Paul says essentially this.

> *Indeed, when Gentiles, who do not have the law, do by nature things required by the law, they are a law for themselves, even though they do not have the law. They show that the requirements of the law are written on their hearts, their consciences also bearing witness, and their thoughts sometimes accusing them and at other times even defending them. (Romans 2:14-15)*

But Paul is referring here to a general sense of right and wrong, not to what constitutes sin in any particular circumstance, and certainly not to all elements of the law of Moses. If it were otherwise, if every individual was preprogrammed with the law of Moses on his heart, there would have been no need for God to give the law through Moses, and no need for Jesus' moral teachings. Neither is Paul here addressing the broader issue that everything that does not come from faith is sin.

Moreover, even a cursory examination of human behavior reveals that most sins are not the result of a conscious defiance of God and the commandments he has given. They are instead the result of ignorance, apathy, impotence, and the like. To a large degree, people sin without being aware that they are doing so.

This is not to say that unintentional sin is not true sin. Leviticus Chapter 4 and Numbers Chapter 15 address unintentional sins at length, and make clear that they incur guilt just as intentional sins do.

> *If anyone sins and does what is forbidden in any of the LORD's commands, even though they do not know it, they are guilty and will be held responsible. (Leviticus 5:17)*

The culpability arises because sinful actions – everything that does not come from faith - perpetuate disorder regardless of whether they were intentional or not. And God, being perfect, cannot abide anything that works against his efforts to bring creation into perfect order. Culpability for unintentional sins also exists because no one is permanently trapped in their sins so as to have an excuse for their behavior. Through faith in Christ's redemptive work, God has provided a way of escape from sin, the punishment for those sins, and from their disordering consequences. Even unintentional sins can be overcome through the sanctifying power of the Holy Spirit which is acquired by that same faith. If someone does not take advantage of this power through faith, they carry the full responsibility for their sin including the punishment for sin.

Paul's statement that everything that does not come from faith is sin also creates a subjective element to the realm of sin: if we act against our conscience, regardless of whether the activity in question is objectively sinful, we have sinned in God's sight. James says exactly this.

> *If anyone, then, knows the good they ought to do and doesn't do it, it is sin for them. (James 4:17)*

Rather than simply stating "it is sin" which would have identified the misdeed as being objectively sinful, James adds "for them" which makes clear that he is speaking of something subjective. Acting against one's conscience is akin to doubting the virtue of our actions. Doubt impedes faith and promotes disorder. It might appear nonsensical to modern western Christians that an action that is not objectively sin could nevertheless still be true sin, but it is consistent with Paul's statement that everything that does not come from faith is sin.

Conclusion

While the modern western church views sin predominantly as intentional rebellion against God and His laws, the Bible treats sin more broadly as any deviation from God's perfect character and perfect ways. Sin is not merely a moral issue, but expresses itself in every aspect of man's existence, including his soundness of mind, his level of maturity, and the health of his relationships. In fact, sin can be described almost entirely in relational terms: man's relationship with himself, his relationships with others, his relationship with creation, and his relationship with God. Insofar as any of these relationships falls short of God's intentions, sin is to blame. While intentional sin is certainly the most egregious form, unintentional sin is still true sin as it both expresses and propagates the disorder that mars life on earth.

With regard to the inclinations within man which lead to sin, there is little in relevant Bible passages to suggest that they are foreign to man as originally created. Instead, the Bible treats such inclinations as innate, something that is the direct result of having been created from earthly materials. It is no wonder that sin is both pervasive and ubiquitous.

Chapter 13

Why Do We Sin?

Under the Traditional Origins Paradigm, the reason that we sin is straightforward: we have a sinful nature that compels us to sin, and this sinful nature was passed on to us from Adam. It was not part of Adam's original makeup but was spawned when he and Eve sinned by disobeying an explicit commandment from God.

The doctrine of original sin, specifically the Augustinian formulation, has been a mainstay of western Christianity for two millennia and there can be no doubt that it meshes admirably well with the doctrine of Christ's atoning work on the cross. With no means whatsoever to escape original sin and thus no means to live a wholly pure and sinless life, man is completely dependent upon God's benevolence and mercy if the relationship between them is to be restored to that which existed between Adam and God prior to the Fall. Even so, there has never been consensus over it among theologians for the church writ large, and there are a host of reasons to question it in light of the relevant Biblical texts. This is in addition to the problem created for the doctrine of original sin by the foundational premise of this book, that Adam and Eve were the products of biological evolution and were not the only people on the earth at the time of the events in the garden of Eden. Adam could not, therefore, have passed a sinful nature to all of humanity as he was the ancestor to only a portion of humanity.

But if the doctrine of original sin is faulty, what then? What would explain man's inability to live a perfectly sinless life? More than that, what would explain man's incessant inclination towards sin, and indeed his capacity for tremendous evil?

Chapter 12 described the breadth of actions and conditions that may rightly fall into the realm of sin, providing some clue as to how sin might arise and proliferate in the natural world absent some form of corruption of the will brought about by Adam's disobedience. This chapter more explicitly connects sin with a humanity that arose through an evolutionary process, and expounds upon the moral state of man in that context.

The core attribute of evolved man

Man is fundamentally self-centered, and this is the inevitable result of the fact that man evolved. More to the point, a focus on self is the key attribute that enabled every single creature to evolve. Every creature that has ever existed has a will to survive that places a priority on its own needs over the needs of others. Every creature seeks sustenance, shelter, and reproductive partners to satisfy its own needs. Every creature seeks to preserve itself, to protect itself, and to advance above the competition. This focus on self is what drives the so-called survival of the fittest, that process whereby those creatures that are best able to meet their own needs are the ones most likely to survive long enough to reproduce and thus pass their genes on to the next generation. Self-centeredness is literally the engine that runs evolution.

This is not to say that there are no examples of selfless behavior in the animal kingdom. A mother rabbit will attack a snake that has gone after its kits when that same rabbit will avoid a snake at any other time. In social communities such as ants and meerkats, individuals operate in ways designed to benefit the group as a whole. In primates there appear occasional examples of altruism wherein an individual takes an action that provides a benefit to another with no obvious or immediate benefit to itself. But these activities merely elevate self-centeredness to the group and species level, a tactic that evolved to ensure their survival over other groups and species. Moreover, selfless behavior does not change the fundamental self-centeredness that drives individuals to place a priority on their own well-being when faced with competition for limited resources and the compulsion to escape from threats.

Man's instinctive self-centeredness makes it nearly impossible for him to be what and whom God wants him to be. The two greatest commandments cited by Jesus in Matthew 22:36-40 - loving God and loving our neighbor - require other-focus. Both sacrifice and obedience require other-focus. Doing good and bearing fruit require other-focus. Man has been programmed by evolution not to do these things.

The Bible implicitly recognizes this natural self-centeredness that drives every single person to seek good for himself. Nowhere does the Bible command that we should eat when we are hungry, rest when we are tired, seek pleasure, spend money on ourselves, spend time with people who are enjoyable to be around, accumulate possessions, and the like. Every person is naturally inclined to do these things, so there is no need to command him to do them.

In addition to placing a priority on his own needs, self-centeredness leads man to trust his own abilities and perspectives over and above anyone else's including God's. It is as if man has tunnel vision, a narrow view of the universe within which he finds it difficult to see anything from a perspective not his own, and even less so to make accommodation for those other perspectives. His understanding of the world around him seems perfectly rational because he does not have access to anyone else's mind; all he knows is what he knows. The same mind that gives him self-awareness is the mind that analyzes, categorizes, and interprets his environment in light of his past experiences. As a result, trusting himself and distrusting others is his default state.

When God imbued our hominid ancestors with an immortal spirit, He began the process of reforming man's outlook and priorities to be aligned with His. That immortal spirit (the image and likeness of God) gave man the ability to connect directly with the Other in a way that was formerly impossible. But having an immortal spirit did not mean that man automatically sought God. Man had the germ of knowledge of the existence of God, but could only know Him by seeking Him, and would only grow in godly character to the degree that he did so. In his *Confessions*, Augustine characterized the state of man's spirit this way:

> *Lord, you have made us for Yourself, and our hearts are restless until they find rest in you.*

Because man's default state was (and is) to trust himself and to place his own needs above all others, seeking God required considerable effort. Worse, it required submission, something that man's self-centeredness was disinclined to do. As a result, man's "restless" heart has gone in search of other things that require less trust and less submission. Deities that may require outward sacrifices but no inward devotion, for instance.

Self-centeredness manifests as sin most obviously in man's aversion to imposed constraints on his behavior. Commandments that are intended to compel him to do this or to refrain from doing that are an affront to his self-determination, and he rebels against them. The so-called lure of forbidden fruit is a good example. The emergence of a prohibition intensifies the desire to do precisely that which is prohibited, as if man seeks to declare that he will be mastered by no one else. Paul identifies this very process at work through the law of Moses.

> *For when we were in the realm of the flesh, the sinful passions aroused by the law were at work in us, so that we bore fruit for death. (Romans 7:5)*

More broadly, any form of imposed constraints - not only moral and spiritual laws, but also civil laws and even social conventions - can elicit a compulsion to nonconformity from self-centered man. Insofar as a person does comply with such laws, in many cases he does so merely out of a desire to avoid negative consequences. If he perceives that negative consequences are unlikely or minimal, he will do as he wishes, or comply in half-measures only. Few comply because they embrace such laws as being for their own good, or for the good of others or of society at large. Fewer still comply for the mere fact that the law was established by an authority figure to whom he is subject. In these rare cases, a person must be of exceptional character or be responding to the prompting of God.

Man's innate self-centeredness is the root of every sin that he could possibly commit, and might reasonably be labelled "original sin."

Name a sin, and you will see a desire to serve self, to feed one's ego, or to place a priority on self over others as the impetus. Conversely, sin is minimized when one's focus is on others. To the degree that self-centeredness outweighs other-centeredness in any word or deed, to that same degree sin will be present.

When a person's focus is the other, he places a priority on meeting their needs even if there is a cost to himself. This is very nearly the definition of love and the very thing that so much of the Bible promotes.

> *it [love] is not self-seeking (1 Corinthians 13:5a, brackets added)*

> *whoever loves others has fulfilled the law (Romans 13:8b)*

To be sure, modern man has a desire to be other-focused, to love as God Himself loves, and this desire is a byproduct of having been made in the image and likeness of God who is defined by love (1 John 4:16). But this godly desire is pitted against his innate self-centeredness which is the byproduct of having emerged from an evolutionary process. In rather personal terms, Paul bemoans this dichotomous state.

> *I have the desire to do what is good, but I cannot carry it out. For I do not do the good I want to do, but the evil I do not want to do - this I keep on doing. (Romans 7:18b-19)*

Through effort man can on occasion be other-focused, and perhaps through monumental effort can ignore his own needs in deference to the other for a time. But his efforts to be other-focused will rarely be completely free from any concern about self, and it is essentially impossible for him to sustain other-focus for a full lifetime. As a result, sin colors every person's life. To be like God in love for others, indeed to love God Himself, and to be free from the sin that self-centeredness spawns, he must rely on God's grace. This requires a full submission to God. In doing so, man is saved not only from the punishment for his sins, but from the compulsion to focus on himself. He is free to trust God to take care of his needs so that he in turn can focus on meeting the needs of others.

The tenacity of sin

Sin has an almost supernatural ability to propagate itself, not unlike a disease as discussed previously. This propagation follows a roughly predictable pattern. Sinful acts lead to harmful consequences; never is it the case that a given sinful act is entirely isolated and benign, having no adverse impact on the person committing the sin or his relationships. The consequences of sin create both internal (man's soul) and external (the world) environments that are more disordered and more out of sync with God. In these environments, man finds that the path of least resistance is to sin again in an attempt to avoid the harmful consequences of previous sins committed by himself or others, or to minimize the discomfort that they cause. It requires greater effort to resist committing a particular sin in an environment already shaped by sin. Thus in addition to our innate self-centeredness, we also sin because in this world it is easier to sin than not to sin.

This pattern of sin propagation means that every person is both a victim of his sinful environment and a perpetrator of sins that maintain and perpetuate that environment. He is a victim in that he is not responsible for creating the circumstances in which he finds himself, and the cumulative effects of his environment over time have the power to injure and warp his soul. But neither is he a purely passive actor in his environment. He chooses to cooperate with his circumstances even when doing so runs afoul of his conscience or of what he knows the proper course to be.

Being a social animal, man wants and usually needs relationships that function well. This is true not only among family and friends, but also within organizations and institutions. Relationships that work well, or are at least tolerably stable, are preferable to relationships that are contentious or apathetic. As a result, every person has an incentive to act and speak in ways that keep those relationships in as good a state as possible, even if those actions and words are sinful. In environments where sin is common and even expected, any attempt to act righteously has the potential to draw criticism or mockery, or to result in being ostracized.

Man's propensity to conform to the prevailing mindsets and activities of those around him is the result of the evolutionary process which gives him an aversion to all forms of discomfort, physical, emotional, psychological, and spiritual. This aversion to discomfort is the natural consequence of self-centeredness and the survival instinct common to all creatures. When given the choice, man will generally choose the less painful path even if that means sinning or continuing to sin.

There are of course occasions in which a person may willingly endure discomfort if he perceives that some greater benefit is possible. Usually this occurs when the benefit is delayed but largely assured. For instance, a person may submit himself to the discomfort of exercise now in order to be healthier later or spend less money now in order to be able to make a larger purchase later. But such actions are still self-centered. Acting in a way that provides a benefit to others rather than to oneself, or in compliance with some moral principle, is much more difficult if there is any discomfort involved. Man remains disinclined to endure discomfort if he sees little or no benefit to himself.

While the natural state of man and the sin-warped state of the world make sinning easy, he is no mere animal, devoid of any knowledge of right and wrong. Man has an immortal spirit made in the image and likeness of God which provides him with a moral compass that always points to true north - the direction of righteousness. But his natural self-centered instincts are often at odds with his God-given moral compass, and the instincts usually win out because they are ingrained in his DNA and precede God's action to give man an immortal spirit. Moreover, man's ability to recognize the direction that his compass is pointing can be impaired by the sinful habits he has formed. The more he listens to and follows his moral compass, the more sensitive he will be to its direction. Conversely, the more he cooperates with his sinful environment, the less sensitive be becomes to his moral compass and the easier it is to continue sinning.

Finally, sin is intractable in the way that it shapes a person's soul through his childhood environment. Before he is capable of making truly independent, rational choices about right and wrong, he has been exposed in his upbringing to views on money, race, authority, sex, and conflict that appear normative. These views influence his developing personhood not unlike the way in which he learns a particular language from his parents. He is thus predisposed to hold the same worldview and repeat the same behaviors as his parents. Since all parents and guardians behave unrighteously in at least some respects, the child they raise will adopt those same unrighteous behaviors. As described in Chapter 12, this creates a multi-generational pattern of sin that is difficult to overcome.

A universe of choices

For most theologians over the history of Christendom, it has been an axiom of the faith that man was created with free will; Augustine and his followers were in the minority in viewing man's free will as being so utterly damaged by Adam's sin that he is incapable of choosing anything but sin. Notwithstanding the doctrine of predestination or its more selective cousin double predestination which in some mysterious way can coexist with free will, God gave man the ability to make any choice he wishes at every moment, on any basis, for good or for ill.

Consistent with man's free will, and doubtless by design, the world presents a nearly infinite variety of options to choose from. This applies in practical ways to where we go and what we spend our time doing. As described in Chapter 11, toddlers know this instinctively - anything they can do they will try to do regardless of whether it is safe or appropriate. The world of options also applies in more morally significant ways to how we treat people, how we respond to laws that apply to us, and our attitudes and activity surrounding sex, power, wealth, and justice.

God gives man free will because it is the only way that love can be authentic. Compelled love is not true love. But He also makes the consequences of our choices clear. Over and over again, in words that could not possibly be any clearer, God describes the blessings available for those who choose rightly and the curses which fall upon those who choose wrongly. He summarizes them this way:

> *This day I call the heavens and the earth as witnesses against you that I have set before you life and death, blessings and curses. Now choose life, so that you and your children may live and that you may love the LORD your God, listen to his voice, and hold fast to him. For the LORD is your life, and he will give you many years in the land he swore to give to your fathers, Abraham, Isaac and Jacob. (Deuteronomy 30:19-20)*

However, by creating a world full of options and giving man the nearly unlimited capacity to choose between them, God consequently and necessarily created an environment wherein godless options are more numerous than godly ones. This is not to say that most options are sinful. More likely, most options are simply morally neutral. But they are not oriented towards God either. Persistent pursuit of godless options leads to ignorance of God, and ignorance of God means disconnection from God. Jesus highlights the fact that options leading to God are in the minority when he says:

> *Enter through the narrow gate. For wide is the gate and broad is the road that leads to destruction, and many enter through it. But small is the gate and narrow the road that leads to life, and only a few find it. (Matthew 7:13-14)*

The range of options to choose from is so great, and the number of righteous and God-oriented options so small in comparison, that few find them and pursue them. Indeed it may be that, without God's help, none would do so. As it has been said, the devil has many faces while God has only one.

Occasionally someone may be cognizant of the constraints involved in being in a right relationship with God. He recognizes that many of the options available to mankind at large are not permissible to the person who has submitted to God and His laws, and he chafes at the affront to his freedom. The Psalmist describes it this way:

> *The kings of the earth rise up and the rulers band together against the LORD and against his anointed, saying, "Let us break their chains and throw off their shackles." (Psalm 2:2-3)*

Once again, man's self-centeredness and the associated claim to self-determination incline him to reject any limitations imposed by God.

Under the Traditional Origins Paradigm, Adam and Eve did not have an infinite variety of choices available to them, many of which led away from God. Instead they had only one determinative choice: whether to obey God's command against eating from the tree of the knowledge of good and evil. Implicit in this view is that every other choice that they could possibly have made, every other option available to them, was a good one. But the Bible says nothing along these lines. It is certainly true that God presented them with only a single command, but the idea that it was impossible for Adam and Eve to sin apart from that command is without any Biblical foundation. Lucifer, the greatest of angels, rebelled against God, and the Bible offers no indication that he did so because of a specific command that he broke. Instead, he became resentful over God's absolute sovereignty. There is no reason to think that Adam and Eve were any different. If they were capable of rebelling against God with regard to the tree of knowledge, they were capable of rebelling against God in countless other ways as well. In this respect they were no different than any other human that preceded or followed them.

The influence of demons

While every person will be held accountable for the choices that he makes, many of the choices which lead to sin are undoubtedly the result of the influence of demons. The Bible makes it clear that Satan and the demons that serve him have been given authority over all the earth. Satan says so explicitly during Jesus' forty days in the desert prior to the beginning of his public ministry.

> *The devil led him up to a high place and showed him in an instant all the kingdoms of the world. And he said to him, "I*

will give you all their authority and splendor; it has been given to me, and I can give it to anyone I want to. (Luke 4:5-6)

Jesus does not dispute Satan's claim to all the kingdoms of the world, and in the gospel of John Jesus refers several times to the "prince of this world." John acknowledges Satan's rule in one of his epistles.

the whole world is under the control of the evil one. (1 John 5:19b)

In the same epistle John identifies demonic activity as the impetus for the incarnation.

The reason the Son of God appeared was to destroy the devil's work. (1 John 3:8b)

The Bible says that demons seek to cause suffering and strife and ultimately to entice people away from God. Peter warns his audience to resist the devil's schemes.

Be alert and of sober mind. Your enemy the devil prowls around like a roaring lion looking for someone to devour. (1 Peter 5:8)

In his letter to the Ephesians, Paul seeks to shift attention away from unrighteous people to the demons that have incited them.

For our struggle is not against flesh and blood, but against the rulers, against the authorities, against the powers of this dark world and against the spiritual forces of evil in the heavenly realms. (Ephesians 6:12)

One can easily imagine unrighteous persons being the unwitting puppets of demons, their sins being in service to Satan's efforts to hinder and destroy God's work.

The Apologists of the 2nd and 3rd Centuries generally saw demons as being the primary cause for evil in the world. According to Kelly, they envisioned demons as

Swarming everywhere, they have obsessed men's souls and bodies, infecting them with vice and corruption.

Tatian and Theophilus saw Adam's act of disobedience as having opened the door to the power of evil spirits for all of his descendants. Similarly, Tertullian admitted to the destructive influence of demons even as he promoted an early form of an inherited inclination to sin. Justin Martyr saw Eve's interaction with the serpent as the first of many demonic deceptions, writing in his *Dialogue with Trypho* of

> *the race of men, who from Adam's time have fallen under death and the deceit of the serpent.*

Notably, it is unclear if Eve would have disobeyed God absent the serpent's deception. The entrance of evil into the world could arguably be attributed to Satan rather than to Eve, though Eve remains responsible for her choice.

Sexual drives

As highlighted previously, Augustine held what today would be considered an extreme view of human sexuality. While asserting that disordered desires of all kinds are expressions of original sin, he focused primarily on sexual desire as the quintessential evil in the heart of man. He viewed all sexual impulses and sexual passion as incompatible with true orientation towards God, even those expressed in the context of marriage. He even went so far as to say that the sexual organs are the very site of original sin within man, and that if there had been no Fall, Adam and Eve would have had children but the sexual act would not have been accompanied by any sexual passion.

Augustine was not alone in his views. Both before and after Augustine, the celibate and ascetic life was considered to be the Christian ideal, the earthly counterpart to life in heaven. Marriage was seen as regrettably necessary for procreation, but tainted by sexual passion and therefore of a lesser quality than the single life. It would not be until the protestant reformation in the 16th Century and the views of such groups as the Puritans that sexuality would come to be seen as good and sexual passion as God-created and approved.

But for all the obsessive theological zeal against sexuality, the early church correctly recognized the power that it has to promote sin.

They contrasted the impulsive and indiscriminate nature of sexual desire with reason and self-control, those attributes that were given by God to set humanity apart from beasts. They also saw in sexual desire not only a distraction from the things of God, but a compulsion to act in ways contrary to God's laws.

These views were not unfounded. When the Jewish believers convened the Jerusalem Council in Acts Chapter 15 to address the question of circumcision for Gentile believers, they concluded that the only elements of the Mosaic law that Gentile believers should be required to follow are certain laws related to food and sexual immorality.

> *You are to abstain from food sacrificed to idols, from blood, from the meat of strangled animals and from sexual immorality. You will do well to avoid these things. Farewell. (Acts 15:29)*

Among the many moral laws that they could have specified, the first leaders of the new church identified sexual immorality as the only one worth mentioning to the Gentile believers. It is evident that they viewed this category of sin as being particularly insidious and thus needing explicit emphasis.

Sexual sins are among those most often targeted by Bible authors, and most often repeated as warnings to their audience. There are twenty-five commandments in Leviticus that describe forbidden sexual relations. Proverbs Chapters 5 through 7 are primarily devoted to sexual sins, with other references to sexual sins scattered throughout the rest of the book. In the major and minor prophetic books, the idolatry of the Israelites is often described figuratively (and sometimes literally) as prostitution or adultery. Paul addresses sexual sins in several his letters, and identifies such sins as representing a uniquely personal form of offense against God.

> *Flee from sexual immorality. All other sins a person commits are outside the body, but whoever sins sexually, sins against their own body. Do you not know that your bodies are temples of the Holy Spirit, who is in you, whom you have received from God? You are not your own; you were bought at a price.*

Therefore honor God with your bodies. (1 Corinthians 6:18-20)

Unrestrained sexual drives are responsible for a wide variety of sins. Those involving some form of violence or intimidation, often resulting in anxiety and trauma, include rape, molestation, pedophilia, coerced prostitution and coerced pornography, and sexual harassment. Even in cases where the sexual activity is entirely voluntary and consensual, it nevertheless has the power to injure relationships between people, disfigure their souls, and impair the relationship between man and God. Such sins include adultery, incest, sex with multiple partners, sex with a prostitute, sex between same-sex couples, and essentially any form of consensual sex outside of marriage. Solitary activities such as watching pornography, masturbation, voyeurism, and compulsive sexual thoughts can also dominate and disturb a person's mind, making it difficult or impossible to make any meaningful connection with God.

Beyond individual sins, sexuality has been at the root of sinful systems and attitudes. Sexism and misogyny were prevalent in Jesus' time - women were seen as weaker than men both physically and spiritually, and were treated as property rather than free agents. They were also seen as being a faulty form of humanity, a deficient version of the perfect male, existing for the purpose of procreation and to serve men. These views of the superiority of males have continued through most of Christian history, fueled by an emphasis upon certain Bible passages such as this:

A woman should learn in quietness and full submission. I do not permit a woman to teach or to assume authority over a man; she must be quiet. For Adam was formed first, then Eve. And Adam was not the one deceived; it was the woman who was deceived and became a sinner. But women will be saved through childbearing - if they continue in faith, love and holiness with propriety. (1 Timothy 2:11-15)

Systemic sexism in early Christendom meant that women were often excluded from full participation in the life of the church. Even today these same attitudes exist in less extreme but still intractable ways in many cultures, and the modern church has yet to fully eradicate them.

As for all sins, sexual sins have their root in the evolutionary process that produced mankind. That process was the same for man as it was for the animals, and sexual drives were likewise the same for man's primate ancestors and for other animals. When God commanded the animals He had created to "Be fruitful and increase in number," they did not cognitively assent to His command and then respond in obedience. Instead, God's command was incarnated as sexual desire, and His creatures simply acted upon that desire. It was a good desire and accomplished God's purposes for creatures that were incapable of understanding the connection between sex and procreation.

God did not stipulate that sexual unions among the animals must be monogamous. The result is that, apart from a few rare cases of monogamy (for example, the gray wolf, barn owl, and bald eagle), animals generally mate with whomever they choose from one mating season to the next. But the Bible makes clear that God's ultimate intention for humanity with regard to sex is different than that for other creatures. In one of the few editorials in the creation narrative, the author of Genesis highlights the unique sexual union between a man and a woman.

> *That is why a man leaves his father and mother and is united to his wife, and they become one flesh. (Genesis 2:24)*

In reference to this very passage, Jesus expounds upon God's intention.

> *So they are no longer two, but one flesh. Therefore what God has joined together, let no one separate. (Matthew 19:6)*

A husband and wife are to be exclusive and committed in their sexual relationship for their full lives.

Every command in the Bible is given for the purposes of pushing human behavior in a direction it would not naturally go, and the command for monogamy is no different. Humans evolved from non-sapient primates for whom sexual desire was the driving force behind procreation, and whose sexual activity was no different than for any other animal in the sense that they were naturally inclined to have sexual relations with any partner at any time. But at some point coincident with or after being imbued with an immortal spirit, God called humans to live differently than the animals. They were commanded to exercise their God-given self-control to limit their sexual activity to a single partner with whom they would remain in a committed relationship until death. This command required humans to defy their natural instincts, the sexual desire worked into their biochemistry through the evolutionary process, to instead place it under the intentional rule of the soul.

The sexual drives of man can rightly be labelled as animalistic as they are truly no different than those for the animals. They are primitive and powerful, with no inherent limitations. If they are not placed under control, they almost always lead to sin. Social conventions occasionally impose some level of control, but in general the control must be exerted by each individual in order to be efficacious.

Conclusion

In a very real sense, the deck is stacked against us. Each one of us enters a world beset with a primordial disorder and contorted by the sinful choices of those who have gone before, coloring every aspect of existence and blinding us to God's goodness and indeed His very existence. Our attention is perpetually oriented towards navigating through a plethora of mostly godless options fraught with the dual traps of sorrow and pleasure. Add to this an innate inclination to place a priority on self and a devil who is bent on our destruction, and we are effectively doomed to remain separated from our Creator. We are both victims of the sinful world in which we find ourselves and also perpetrators of sin which keep us and the world separated from God. God Himself must intervene if there is to be any hope for mankind.

It could have been different. Had Adam and Eve remained faithful to God and their descendants followed suit, we would each enter a world where God's will reigned and where the knowledge of God was commonplace. The devil would presumably have been evicted from the earth, and humankind would be reigning as God intended. Life for all would be something closer to what the children of strong Christian parents experience currently, but extended beyond their family to all environments.

But even if that were the case, we would still have the innate self-centeredness that is the result of having emerged from an evolutionary process. This would leave us in the same state as Adam and Eve before the Fall, requiring constant vigilance to trust in God over self. In effect, the test that Adam and Eve faced would ever remain with us, or at least so long as we lived on earth. This is the peculiar world that God created for His most favored creatures, a place where glory comes to those who triumph through intentional dependence upon God and His grace.

Part IV

The Plight of Man

Chapter 14
A New Paradigm

Correctly identifying and explicating Christian truth is a messy process. While Christians have the advantage of the indwelling Holy Spirit who, according to Jesus, "will teach you all things" (John 14:26), in practice there is much consideration and debate involved, and often balloting. Doctrinal views of the Trinity, the dual nature of the Christ, baptism, the Eucharist, and the nature of man before and after the Fall all took centuries to take their current form, having first gone through various permutations with conflicting views being commonplace. Some of the earliest beliefs are now considered askew or outright false. A disquieting example is the church's rigid stance on geocentrism which was not corrected until the 18th Century.

Doctrine is not immutable. It can and does change. In some Christian denominations this fact is more accepted than in others. In those denominations where the views of church fathers are more highly esteemed or where doctrines and dogma are formally codified, changes in doctrine are permitted only insofar as they can be described as developments of pre-existing doctrines or as an expansion of understanding that does not alter any of the original underlying precepts. But even in denominations where changes to doctrine may be more readily accepted, Christians are right to maintain a skeptical attitude and to proceed cautiously when presented with any new concept. Long-standing interpretations of the Bible should always be treated with respect until and unless there is a compelling reason to change them.

The Revised Origins Paradigm is an attempt to explain the universal sinfulness of mankind in light of the relatively recent scientific evidence for biological evolution and the unavoidable consequence that Adam and Eve were not the only humans on the earth during the events of the garden of Eden. But the Revised Origins

Paradigm does not question the underlying premise that every human being is a sinner, born with an inclination towards sin. Indeed, it acknowledges this as an observable fact. Further, the Revised Origins Paradigm retains that essential theological requirement that all of mankind is in need of a Savior. As a consequence, it need not be viewed as a new doctrine, but rather a development of an existing doctrine.

Even so, the Revised Origins Paradigm will undoubtedly alarm traditionalists because it must necessarily address a host of other issues, including how the narrative in Genesis Chapters 1-3 could fit with a humanity that emerged out of an evolutionary process and assumptions about suffering and death. Treating some portions of that narrative as figurative or symbolic while retaining others as literal is not new - even the Traditional Origins Paradigm does this - but the particulars presented in this work are nevertheless a departure from views held by many (though certainly not all) Christians since at least the 5th Century.

Insofar as one accepts biological evolution as fact, or even as a legitimate possibility, the Revised Origins Paradigm represents a reasonable path towards maintaining the core doctrines of the Christian faith. What follows is a summary of those elements of the Revised Origins Paradigm that figure most prominently into its credibility, and the reasons why any Christian can accept it and be assured of their ongoing good standing with God.

The story of Adam and Eve

Unlike some approaches which treat the Genesis account of the affairs of Adam and Eve as mostly if not entirely figurative, the Revised Origins Paradigm affirms the literal truthfulness of several features of the traditional view:

- Adam and Eve were real people
- The garden of Eden was a real geographical place

- God intended Adam and Eve to be His regents on the earth to spread His influence and reign over creation
- The tree of life and the tree of the knowledge of good and evil were real trees
- God gave an explicit command to Adam that he was not to eat from the tree of the knowledge of good and evil
- The devil deceived Eve into eating from the tree of the knowledge of good and evil, and Adam also ate of it
- God expelled Adam and Eve from the garden of Eden for their disobedience
- The sin of Adam and Eve resulted in disastrous consequences for all humanity, namely that sin and spiritual death proliferated

The Revised Origins Paradigm also maintains the truthfulness of the Genesis account of the creation of the physical universe, the earth, all life, and humanity in its essential features, namely that it involved a process guided by God. It deviates from the Traditional Origins Paradigm in that it does not hold to a literal six-day creation period but rather assumes an evolutionary process spanning billions of years. Suffering and death were an inherent aspect of this process, in contrast to the traditional view under which the world was created perfect and contained no carnivores, no accidents, no storms, and no pain or discomfort of any kind.

As for creation as a whole, Adam and Eve were likewise not perfect, morally or otherwise. They were special in that God chose them for an important task, and undoubtedly were of sufficiently robust character that they were capable of carrying out that task. But they were flawed, just as was every other human being at that time. They were also mortal, destined to die unless they ate from the tree of life.

The most prominent departure from the Traditional Origins Paradigm regards the precise nature of the Fall and its impact on humanity. By mistrusting God's goodness and providence, Adam and

Eve did what they were explicitly commanded not to do. As a result, they were deprived of access to the tree of life. They lost their opportunity to become immortal, and as God's chosen representatives on earth they also lost that opportunity for the rest of humanity. As a result,

> *in Adam all die (1 Corinthians 15:22a)*

God had intended Adam and Eve to live a life of ample provision and joyful work serving Him in the garden of Eden. But being cut off from the tree of life meant also being cut off from the bounteous garden, and they returned to the impoverished outside world where their efforts to sustain themselves would require considerably greater effort. There they would be forced to trust in God to care for them in ways that were less necessary in the garden of Eden.

When Adam and Eve disobeyed God, they became sinners. However, their core natures did not change - they were already imperfect and self-centered. They became sinners by the simple fact that they had sinned by breaking the command of God. They fell into a new category before God, no different than if they had broken a thousand commands.

> *For whoever keeps the whole law and yet stumbles at just one point is guilty of breaking all of it. (James 2:10)*

Adam and Eve had not sinned prior to their life in the garden of Eden because there had previously been no command of God to break, and

> *where there is no law there is no transgression (Romans 4:15b)*

Since God's command to Adam and Eve was the first command to have been given to humanity, it was the first opportunity for any party to overtly sin, though being flawed human beings they had certainly fallen short of moral perfection prior to the garden of Eden and had acted, spoke, and thought in unrighteous ways.

After their expulsion from the garden of Eden, Adam and Eve as well as their children continued to have a relationship with God, albeit with differing levels of devotion. But the moral debasement which characterized the rest of humanity soon ensnared them as well, and over subsequent generations this relationship with God withered away to essentially nothing. Though Adam had been intended by God to be the locus of a new era of righteousness and faithfulness to God on the earth, his failure ignited a domino effect of sin, affecting all his progeny.

> *through the disobedience of the one man the many were made sinners (Romans 5:19a)*

Sin proliferated by its tenacity (the world) and through man's innate self-centeredness (the flesh), aided by the work of demons to beguile and tempt mankind.

The process that God used to bring humanity into existence and to begin His relationship with them under the Revised Origins Paradigm may appear insufficiently supernatural to some, an insinuation of deep time and biological evolution into the simpler traditional story of a near-instant creation of a perfect world. But little about reality is truly simple, and mysteries abound in the Christian faith. How and why God would use such a process is in some respects no different than the incomprehensible nature of the Trinity, or the coexistence of predestination and free will, or the dual nature of the Christ, or suffering among the innocent and God-fearing. We can describe these mysteries but not truly explain them; we can accept them but are forced to relegate their full reality to the inscrutable God.

Even so, Paul provides a possible clue to God's purposes in his speech to the Athenians. In consideration of their pagan culture and preoccupation with philosophy, he describes God's purposes in creating humanity and distributing them around the earth as being done in the hope that they would recognize His work for and among them and thereby seek and follow Him.

> *God did this so that they would seek him and perhaps reach out for him and find him, though he is not far from any one of us. (Acts 17:27)*

The phrase "reach out" can be translated as "feel their way" or even "grope." One imagines being lost in the dark and moving towards the sound of help in the distance. God's plan for humanity from the beginning of time has been that they would respond to the spark of divinity that He placed within them and the revelation of His glory in creation to make a meaningful connection with Him. An evolutionary process for the emergence of humanity that spans millions of years might seem excessively long and even wasteful, but it is consistent with Paul's description of God's purposes.

Order out of disorder

In Genesis Chapter 1, God is shown going through a process of ordering the world. He separates light from darkness to establish diurnal cycles, separates waters to create earth and sky, and moves the waters on earth to permit dry ground to appear. He demonstrates that the chaos of the natural world is no match for His power as He brings it under His control and organizes it to suit His purposes. Centuries later, the Psalmist highlights this same sort of activity in the context of recounting His awesome attributes.

> *[God] stilled the roaring of the seas, the roaring of their waves, and the turmoil of the nations. (Psalm 65:7, brackets added)*

Notably, the Psalmist places God's work to bring peace and order to creation alongside His work to bring peace and order to humanity. In one sense, it is one and the same work for God. The Bible provides a similar pairing in Jesus. He gives peace to His followers in the midst of the chaos of the world, but He also rebukes the chaos that exists in creation.

> *Peace I leave with you; my peace I give you. I do not give to you as the world gives. Do not let your hearts be troubled and do not be afraid. (John 14:27)*

> *He got up, rebuked the wind and said to the waves, "Quiet! Be still!" Then the wind died down and it was completely calm. He said to his disciples, "Why are you so afraid? Do you still have no faith?" They were terrified and asked each other,*

> *"Who is this? Even the wind and the waves obey him!" (Mark 4:39-41)*

The disciples' reactions indicate that, though they had seen Jesus do a variety of miracles before, commanding the wind and waves to be still was an entirely new category of miracle, the infinite Almighty taking control over the chaos of nature. It was another step in their burgeoning awareness that Jesus was not merely sent by God, but was God Himself.

Although God began the process of bringing order out of disorder at the beginning of time, that process did not result in an idyllic world free from all disorder. God passed the baton to Adam and Eve and commanded them to subdue the earth and rule over it. There would have been no need to subdue anything if it was already in perfect order. Even the garden of Eden, as full as it was of every resource that Adam and Eve needed, still required effort to organize it; Adam was charged with working the ground to cultivate and care for it, something that would have been unnecessary if it was already perfectly suited to provide all their needs. The world outside the garden, to which Adam and Even would eventually be exiled, was even less suitable and would require greater effort to provide the sustenance they needed.

God's process of bringing order out of disorder, and mankind's enlistment into this endeavor, is present as an undertone throughout the Bible. From the beginning it has been God's plan to bring shalom to the world in human hearts, their relationships between one another, their relationship to God, and in the natural world. Shalom is a comprehensive form of peace encompassing not just personal contentment and tranquility, but also completeness, health, welfare, safety, prosperity, peaceable friendships, and harmony between nations. It is the ultimate form of order and harmony that can only occur when all of creation is brought into right relationship with and subjection to God. Paul describes this crowning purpose as one which is *to be put into effect when the times reach their fulfillment - to bring unity to all things in heaven and on earth under Christ. (Ephesians 1:10)*

God's efforts to bring order out of disorder can be seen in the charges that He gives to His followers. He organizes civil life by establishing various earthly authorities and commands us to live in subjection to them.

Submit yourselves for the Lord's sake to every human authority (1 Peter 2:13a)

He brings harmony into personal relationships by commanding His followers to promote peace in the face of strife, even labeling peacemakers as children of God.

If it is possible, as far as it depends on you, live at peace with everyone. (Romans 12:18)

Matthew Chapter 5 provides a selection of Jesus' prescriptions for right relationships, all of which are designed to promote order and harmony just as much as they fulfill righteousness. After emphasizing the ongoing relevance of the Mosaic law which is replete with commandments that order Jewish life, He highlights and expounds upon the proper response to disagreements and anger, insults and unreasonable demands, and the importance of honesty in all relationships. Much of this chapter is considered to be an elaboration of the royal law "Love your neighbor as yourself" cited in both Galatians 5:14 and James 2:8, both of which repeat the command given in Leviticus 19:18. Not only does this application of love neutralize the self-centeredness that is a core attribute of evolved man, but order and harmony are promoted when there is consistency between how we treat others and how we treat ourselves.

Sin is the most consequential expression of disorder in the universe as it leads to disequilibrium in and disruption of community, leading to disconnection. James provides a direct link between sin and disorder.

For where you have envy and selfish ambition, there you find disorder and every evil practice. (James 3:16)

Martin Luther viewed the primary source of man's sinfulness as disharmony and disorder in one's soul, which in turn derived from "the radical and fundamental self-centeredness which characterizes all our actions" (according to Alfred Vanneste, writing of Luther's views).

There is no disorder or disharmony within God, nor in His dwelling place, nor in and between those creatures that are near to Him. But disorder does arise in God's creation. When this happens, God takes action to address it. How He does this, and over what timeframe, is uniquely His to decide and His wisdom in these matters is inscrutable.

When Lucifer rebelled against God, for a moment there was disharmony in heaven. In that case, God acted swiftly to exclude him and his followers from His presence. But God did not annihilate Lucifer, nor did He immediately consign him to hell. For some unfathomable reason, God gave Lucifer freedom both to roam the earth and to act to sow disorder and disharmony. God decided that allowing some disorder was a fitting element of His plan for creation.

God takes the same approach with mankind: those who intentionally promote disorder are left largely unhindered. To the dismay of many, even those in positions of authority and influence, be they parents, teachers, police, employers, or governing officials, are not immediately punished or removed from their positions when they oppress or abuse those under their control. Instead, God enlists His own servants to rectify the disorder that has been wrought. As at the beginning of creation, mankind is God's partner in bringing order out of disorder.

Imperfection is intrinsic to creation, and where imperfection exists there is at least the potential for disorder, if not the inevitability. Under the Traditional Origins Paradigm, all that exists must have been created perfect in its original state because it was derived from the perfect God. But exactly because it is derivative, creation must in reality fall short of God's perfection in some fashion; creation may be of God, but it is not God. In one of his speeches to Job, Eliphaz makes this very point.

If God places no trust in his holy ones, if even the heavens are not pure in his eyes, how much less mortals, who are vile and corrupt, who drink up evil like water! (Job 15:15-16)

Under the Revised Origins Paradigm, the disorder that existed in nature at the beginning of time, which God began to address before turning the task over to Adam and Eve, also existed in man's soul. Paradoxically, God enlisted imperfect humans to help bring order to an imperfect creation. Insofar as Adam and Eve remained in unity with and subjection to God, they could accomplish this task because God would give them everything they needed, materially, mentally, emotionally, and spiritually. Apart from God, their own abilities were insufficient. After their disobedience sullied their relationship with God, this was proven true as they drifted into reliance on self and their efforts to bring order out of disorder faltered and eventually evaporated. Humanity by and large then became agents of disorder rather than of order.

Transcending man's animal heritage

As creatures that came to be through an evolutionary process, human beings carry within them the natural instincts that helped their less-sentient and non-sentient ancestors to survive and which are common to all animals. Self-centeredness is the primary such instinct. Self-centeredness drives each individual to pursue its own well-being above the needs of others. Those that are most successful in this endeavor are most likely to reproduce and pass their genes on to the next generation. The stronger almost never give heed to the weaker; there is no sense of fairness in the animal kingdom, no compulsion on the part of any individual to ensure that all are given equal access to food and shelter or equal opportunities to mate.

By virtue of the immortal spirit implanted into human beings at some relatively recent point in their evolutionary history, they have within them the image and likeness of God. And as God is other-focused, so also is the immortal spirit of man. But this spiritual instinct is at war with man's natural self-centeredness, ingrained within his DNA. Unless he is in relationship with God, man has only a limited ability to be other-focused. More specifically, man cannot be consistently and thoroughly other-focused without the abiding presence of God to encourage and empower him to be so.

It is God's intention for every human being that he transcend and conquer the animal instincts that are his natural heritage; according to Blaise Pascal, the future destiny of every person is either divinity or bestiality. In becoming other-focused, man learns to love his enemies, forgive those who hurt him, and give sacrificially to those in need. He also learns to control his appetites, practice delayed gratification, and trust God in difficult circumstances rather than railing and struggling to manipulate them to his advantage. Whatever the normal and common mode of behavior is for animals, most notably our closest biological cousins chimpanzees and gorillas, man is meant to renounce and replace with the fruit of the Spirit: love, joy, peace, forbearance, kindness, goodness, faithfulness, gentleness, and self-control.

The Bible likens sinful behaviors in humans to the natural instincts of animals. In some cases this comparison is explicit.

> *But these people blaspheme in matters they do not understand. They are like unreasoning animals, creatures of instinct, born only to be caught and destroyed, and like animals they too will perish. (2 Peter 2:12)*

In other cases the comparison is implicit.

> *These are the people who divide you, who follow mere natural instincts and do not have the Spirit. (Jude 19)*

In this verse, "natural instincts" derives from the Greek word *psuchikos* which generally means breath in the sense of the bodily animation which men have in common with animals. In Jude it refers more specifically to man's sensuous, animal nature with its subjection to appetite and passion. James also uses this word, connecting it both to the earth and to the demonic.

> *Such "wisdom" does not come down from heaven but is earthly, unspiritual [psuchikos], demonic. (James 3:15, brackets added)*

There are also multiple references in the Bible to the fact that man is made of earth-stuff, and in that respect is no different than animals. The Greek word *sarx*, commonly translated as flesh, is used in reference to both man and animals to describe their physical bodies. It is also used to describe the sensuous nature of man which is the same as that for animals, and which incites man to sin.

> *For when we were in the realm of the flesh [sarx], the sinful passions aroused by the law were at work in us, so that we bore fruit for death. (Romans 7:5, brackets added)*

> *For if you live according to the flesh [sarx], you will die; but if by the Spirit you put to death the misdeeds of the body, you will live. (Romans 8:13, brackets added)*

Sarx is sometimes translated as sinful nature rather than flesh, making the connection between man's physical body and sin that much clearer.

> *For I know that good itself does not dwell in me, that is, in my sinful nature [sarx]. (Romans 7:18a, brackets added)*

Paul also uses the phrase "earthly nature" to describe the source of sin.

> *Put to death, therefore, whatever belongs to your earthly nature: sexual immorality, impurity, lust, evil desires and greed, which is idolatry. (Colossians 3:5)*

Man's physical body, having been produced through an evolutionary process no different than that for other animals, is subject to the same natural inclinations as animals, predominantly self-centeredness. When man acts as animals do, he generally acts contrary to the precept of other-centered love that is God's defining attribute. According to Wiley, even Thomas Aquinas recognized that sin is a problem of nature rather than an exclusively moral problem. While biological evolution was not under consideration in the 13th Century, Thomas nevertheless pointed to the limitations of man's natural capacities as both the locus of man's sinfulness and his need for the active and invigorating power of God to incline man towards righteousness.

Original sin

In common Christian vernacular, original sin means simply an intrinsic inclination towards sin that every human being is born with and which colors all of his thoughts, words, and actions. By this definition, the Revised Origins Paradigm is in agreement with the Traditional Origins Paradigm. The difference between the two paradigms lies in the proposed source or cause of that sinful inclination, and its implications for the other elements of the formal doctrine of original sin as put forth by Augustine.

By virtue of his evolutionary heritage, man is predominantly inclined to serve himself, to prioritize his own needs over the needs of others. He will generally seek to gain advantage for himself, to gather resources for himself, and to prioritize pleasure and comfort for himself while downplaying or dismissing the consequences for others. There may be occasions in which unredeemed man is other-focused, for instance in close familial relationships or where he is dependent upon another for his well-being. But this is generally the exception rather than the rule.

There is no escaping this innate self-centeredness and the sin that it incites. Every person without exception inherits it from his parents, and in this sense it is no different than under the Traditional Origins Paradigm. But unlike the traditional view, this inheritance is not the result of a change in Adam's nature which occurred when he disobeyed God. Indeed there is nothing in the Bible that explicitly describes any such change in Adam's nature. Instead, it is an inference drawn from the assumption, again not explicitly stated in the Bible, that Adam was created morally perfect, combined with Paul's statement that

> *through the disobedience of the one man the many were made sinners (Romans 5:19a)*

In the traditional view, Adam could only have made many other sinners if he passed a sinful nature on to them through procreation, and he could only have done that if he himself had such a nature. As he was created perfect, ipso facto his nature must have changed when he disobeyed God.

Under the Revised Origins Paradigm, Adam was already imperfect and self-centered when he entered the garden of Eden, and he disobeyed God by mistrusting God and trusting himself. In so doing, Adam's sin set an example for his progeny. Not only that, but the primary consequence of his sin - exclusion from the garden - also adversely impacted them. Without the special protection and abundant provision provided in the garden, and with the promise of immortality gone, Adam's descendants found themselves following in Adam's footsteps. Rather than trusting God to provide for them, they trusted in their own wisdom and abilities, eventually so much so that they forgot God altogether. God had intended Adam to set a positive example for his progeny and all the rest of contemporary humanity for what it meant to be in proper relationship with God. Instead he failed in his appointed task and thereby set a negative example, interrupting God's plan to bring all of humanity to Himself.

Under Augustine's formulation of original sin, man's inclination towards sin is utterly irresistible. Unredeemed man can do nothing but sin. He can never choose the good, and therefore cannot please God at any time. He is thoroughly and forever a sinner in every aspect of his being, and any appearance of righteousness is merely pretense masking impure motives. Under the Revised Origins Paradigm, this is not the case. By virtue of the immortal spirit within, every human being has been created in the image and likeness of God and so has the capacity to act righteously. But because man is also the product of an evolutionary process, he simultaneously has a propensity towards self. Man is not incapable of resisting sin and doing good, but it takes effort and a conscious act of the will. His natural state and the path of least resistance is to serve himself above others and above God. A lifetime of fully righteous acts, thoughts, and words is essentially impossible.

Augustine's formulation also asserts that man is guilty not only of his own sin, but also of Adam's sin. Although an infant does not have the capacity to make reasoned choices and therefore cannot intentionally sin, it is nevertheless a sinner and deserving of eternal punishment. This guilt derives from seminal presence, the concept that every human being was in Adam's body when he sinned, and therefore acted with him. Seminal presence, in turn, relies on the premise that Adam and Eve were the progenitors of all of humanity.

But because Adam and Eve were in reality the products of evolution, they were not the only humans on earth at the time of the events in the garden of Eden. As a result, humanity as a whole cannot be said to have been seminally present with Adam when he sinned and cannot have acted with him. Insofar as Adam was appointed by God to represent all of humanity, it is possible that Adam's guilt is imputed to all others. Regardless, it is certain that all of humanity suffers the punishment for that guilt, physical death, by being deprived of access to the tree of life.

The advantages of the Revised Origins Paradigm

The Revised Origins Paradigm is more effective than the Traditional Origins Paradigm at explaining the current state of affairs with regard to the natural world and man's spiritual position before God, and the process that led to them. To begin with, traditionalists, who are typically young-earth creationists, must contend with the science of radiometric dating of rocks which indicates that the earth is 4.5 billion years old and with the fossil record which demonstrates that life on earth began with simple organisms and progressed over time to more and more complex ones. They must also contend with fossil discoveries laying out a pattern of development of bipedal hominids which ultimately led to homo sapiens. Because these facts do not comport with a more literal reading of Genesis Chapters 1 and 2 under which the earth was created about 6,000 years ago and every creature and plant came into existence within the span of six days, many traditionalists decry the science itself as being at best erroneous and at worst the work of Satan. Some go so far as to insist that the traditional view is the only acceptable one, with any alternative views being tantamount to rejection of the truth of God's Word and thus God Himself. In their zeal to maintain fidelity to Scripture, they can actually drive spiritual inquirers away from the Christian faith.

More pertinent to the focus of this book, the Traditional Origins Paradigm leads to a number of dilemmas regarding Adam and Eve's sin and the effect that it had not only on all of humanity but also on all of creation. For instance, the mere fact that they sinned is in direct conflict with their purportedly perfect natures; how does one sin if one cannot conceive of sin? The Revised Origins Paradigm provides a straightforward answer to this dilemma: Adam and Eve were born with a natural inclination to prioritize their own desires and needs over all else, and to mistrust others by default. This self-centeredness was the result of the evolutionary process that produced them, and it was critical to the survival and advancement of their more primitive ancestors.

The Traditional Origins Paradigm asserts that Adam's moral nature changed when he sinned, but has struggled to describe that change with any precision. Most critically, how could a single poor choice have changed the nature of Adam's soul from thoroughly good to thoroughly evil? Early theologians who agreed with the Augustinian formulation of original sin were certain that such a change must have occurred because of their adherence to the premise that Adam was created perfect, but they were forced to speculate on the nature of that change. Some described the change in Adam's soul as something evil added to it, a stain or contagion. Others described it as the removal of something good such as "original justice" (per Anselm of Canterbury), or even the withdrawal of the Holy Spirit. Still others saw it as a contorting of his soul, damage to its machinery that impaired its proper functioning. But none of these speculations directly addressed how it occurred. They were merely descriptive, not explanatory.

Early theologians also found themselves speculating about how Adam's newly acquired sinful nature could have been passed on to his progeny. There was general acceptance that it occurred through the procreative act, but how the eternal soul of the child was connected to that of its parent was a mystery. Two concepts arose to fill the void: traducianism and creationism (no relation to the much later use of the same word to describe the origin of the universe and life on earth).

Traducianism proffered that an individual's soul is derived from the souls of its parents. In essence, there is a budding of sorts, a splitting off of a portion of the parent's soul that becomes the soul of the child. As the moral orientation of an individual is contained within his soul, the child receives not only his soul from his parents but also his moral nature. The primary support for traducianism is found in this verse:

> *When Adam had lived 130 years, he had a son in his own likeness, in his own image; and he named him Seth. (Genesis 5:3)*

Supporters of traducianism saw a parallel between God creating Adam in God's own image and likeness, and Adam procreating Seth in Adam's image and likeness. God was morally perfect and so passed His perfection to Adam. Adam was not morally perfect and so passed his imperfection to Seth. Traducianism was built upon the concept of seminal presence under which every individual that would ever exist was present with Adam when he sinned and thereby acted with him. In so doing, every individual's eternal soul was marred by sin at the moment that Adam sinned. Traducianism was favored by most Latin Christians including Augustine as it provided a basis for the inheritance of original sin.

There was strong opposition to traducianism from predominantly Greek Christians. In addition to taking exception to the idea that the procreative act could pass the immaterial from one generation to the next along with the material, they also condemned the idea that it is man rather than God who effectively creates, and chooses to create, a new soul. They cited Bible verses supporting the alternative concept of creationism under which God creates each individual soul Himself ex nihilo (from nothing) and implants it into a fetus's body.

> *From heaven the LORD looks down and sees all mankind; from his dwelling place he watches all who live on earth - he who forms the hearts of all, who considers everything they do. (Psalm 33:13-15)*

> *This is what God the LORD says - the Creator of the heavens, who stretches them out, who spreads out the earth with all that springs from it, who gives breath to its people, and life to those who walk on it (Isaiah 42:5)*

Unsurprisingly, creationism did not comport well with Augustine's formulation of original sin. How could a sinful nature be transmitted from one generation to the next if the eternal soul, the repository of that sinful nature, is created directly by God?

Traducianism carried with it other problems that were left unresolved. How were the "budded souls" of both parents combined to produce a single soul for the child? How could the moral nature of the parent be transmitted to the child without also transmitting the parent's personality or memories? How could an immaterial soul be located in and transmitted with sperm and/or egg?

The Revised Origins Paradigm eliminates the need to speculate on how on sinful nature could be transmitted from parent to child. Man's sinful nature is in reality just his inclination towards self that is the natural, necessary, and inevitable result of his having emerged from an evolutionary process. Every person inherits this self-centeredness from his parents in the same way that he inherits two arms and two legs; it is an innate part of his personhood, something that no person can be born without.

The Revised Origins Paradigm also provides a straightforward way to explain the existence of disorder and savagery in the natural world - it is the result of the process of natural selection in which all plants and animals are instinctively driven to gain an advantage over their competitors in order to survive, and those that are best able to do so are positioned to pass their genes on to the next generation. It is this process that creates predator-prey relationships, spines and tusks, venomous teeth and stingers, poisonous skin, camouflage and mimicry, and claws and teeth specially designed to tear flesh. The Traditional Origins Paradigm has little to say about how the sin of Adam could have changed the lion's teeth and biochemistry from a plant-based diet to a meat-based diet, or could have created carnivorous plants and deadly viruses, or could have changed the very nature of physical reality to permit accidents in a world that had previously been perfect and free from every possible source of suffering.

Finally, there is the conundrum of the ongoing reality of physical death in light of the fact that Jesus's work on the cross eliminated death for the believer. On one occasion Jesus Himself acknowledged that every Christian still physically dies before being resurrected in a new body.

> *Jesus said to her, "I am the resurrection and the life. The one who believes in me will live, even though they die" (John 11:25)*

But on other occasions Jesus's words blurred the distinction between physical death and spiritual death.

> *Your ancestors ate the manna in the wilderness, yet they died. But here is the bread that comes down from heaven, which anyone may eat and not die. (John 6:49-50)*

> *"Very truly I tell you, whoever obeys my word will never see death." At this they exclaimed, "Now we know that you are demon-possessed! Abraham died and so did the prophets, yet you say that whoever obeys your word will never taste death." (John 8:51-52)*

> *whoever lives by believing in me will never die. (John 11:26a)*

It appears that Jesus saw any physical death which was followed by resurrection into eternal life with God as being equivalent to escaping physical death altogether. Generally, He viewed true death as referring only to spiritual death (separation from God and the disorder that results from such separation).

The believer's escape from punishment through the atoning work of Christ on the cross is the central doctrine of Christianity, and there is no dispute over the fact that the believer becomes a new creation immediately upon belief in that atoning work. There is no waiting period, no need to arrive in heaven before his justified position before God is activated. The difficulty arises when traditionalists insist that God's punishment for sin is both spiritual death and physical death, after their interpretation of God's warning to Adam.

> *you must not eat from the tree of the knowledge of good and evil, for when you eat from it you will certainly die. (Genesis 2:17)*

If the punishment for sin is both spiritual death and physical death, and that punishment was transferred from man to Jesus on the cross, the believer should become a new creation both spiritually and physically. Physical death should be removed immediately just as spiritual death is removed immediately; there should be no physical death for the believer. The fact that believers do physically die is evidence that the punishment for sin is not physical death but rather spiritual death. The only alternative under the Traditional Origins Paradigm would be that Jesus's work on the cross was not meant to remove the punishment of physical death even though it did remove the punishment of spiritual death. Effectively, the believer retains one form of punishment while escaping the other. This outcome is nonsensical, not to mention unbiblical.

The Traditional Origins Paradigm attempts to resolve this by emphasizing that, though every person physically dies, physical death is not ultimately consequential for the believer. It is merely a passage from earthly life to life in heaven, with the attendant transformation of the mortal body into an immortal body. Paul describes this very thing.

> *So will it be with the resurrection of the dead. The body that is sown is perishable, it is raised imperishable; it is sown in dishonor, it is raised in glory; it is sown in weakness, it is raised in power; it is sown a natural body, it is raised a spiritual body. If there is a natural body, there is also a spiritual body. (1 Corinthians 15:42-44)*

While this is certainly true, it does not address the question of why the believer should physically die at all. Moreover, it is not as if physical death were "merely" a passage. Physical death is an ugly process, typified by weakness, incapacity, and pain often drawn out over a long period of time. Surely the punishment of physical death, if it is indeed a punishment, includes these unsavory elements, and the cross does nothing to remove them.

Under the Revised Origins Paradigm, Adam and Eve were created mortal and so were destined to physically die. So long as they were in the garden of Eden and enjoyed God's favor, they had access to the tree of life whose fruit forestalled physical death. But when they were expelled from the garden their physical bodies reverted to the aging process and susceptibility to injury and disease that is common to all animals. Although in one sense their punishment was (among other things) physical death insofar as immortality was no longer available, in another sense physical death was not a punishment per se but rather the indirect result of their being cut off from access to the tree of life. The true death was spiritual, separation from God and being confined to a state of unrighteousness before Him.

Not only that, but one might say that death was not a punishment at all, but merely a consequence of their sin. That is, it may be that God did not assign death, spiritual or physical, to Adam and Eve following their disobedience, but instead let that disobedience have its natural and inevitable end. Although God's warning to Adam prior to the Fall clearly indicated that death would be the result of eating from the tree of the knowledge of good and evil, He did not say that He would impose death on Adam, but only that Adam would die. After the Fall when God pronounced His judgments, He did not mention death as one of the punishments. Instead, He said only that

> *By the sweat of your brow you will eat your food until you return to the ground, since from it you were taken; for dust you are and to dust you will return. (Genesis 3:19)*

Hard toil was the punishment. Physical death was Adam's natural destiny.

So, then, the work of Jesus on the cross was aimed at atoning for our sins and removing the punishment of spiritual death, restoring our right relationship with God. It did not remove the punishment of physical death because physical death was not a punishment imposed on man for his sins. Jesus' resurrection provided the remedy for physical death, though not its elimination.

Unity despite doctrinal differences

There will undoubtedly be some who, while perhaps not disagreeing with the rationale behind the Revised Origins Paradigm, are nevertheless wary of subscribing to a doctrinal position that deviates from more familiar and long-standing views. This is both understandable and healthy, a sign of a desire to be faithful to God and to respect the conclusions of the experts who have also studied this topic.

At the same time, it is important to consider the enormous variety of doctrine that Christians have held over the last two millennia and even today, and to place the Revised Origins Paradigm into that context. One has only to look at the enormous number of Christian denominations to appreciate this doctrinal diversity. Even putting aside religions that do not pass the test of Christian orthodoxy such as the Church of Jesus Christ of Latter Day Saints, Jehovah's Witnesses, and the Unification Church, there are around 200 distinct denominations in the United States alone. Each new denomination typically arose due to disagreements with already established denominations over doctrine, though in some cases the driving issues were related to authority or administration. It is safe to say that every denomination believes that their doctrine is superior to and more soundly Biblical than every other denomination's doctrine. If it were not so, there would have been no need (or more correctly, no motivation) to form a new denomination.

Much of the variety in doctrine surrounds legitimately disputable issues that do not impinge upon the core truths of the Christian faith. Such core truths are often termed dogma and represent those indispensable doctrines that must be correctly understood and believed in order to have right standing before God. At their center are those directly related to salvation: man's inescapably sinful condition before God and his consequent subjection to God's wrath, God's gracious provision for man's helpless condition through the incarnation, man's justification before God through faith in Christ's atoning work on the cross, Jesus' bodily resurrection, and redeemed man's eventual resurrection into a new body and his eternal residence with God in heaven. Many extend the core doctrines to include such things as the Trinity, the virgin birth, Jesus was sinless and was simultaneously fully God and fully man, the Second Coming, the Bible is the inerrant word of God, and God is the creator of all that exists.

The core doctrines distinguish Christianity from all other religions, and there can be no disagreement between orthodox Christians about them. As a result, all true heresies involve some deviation from core doctrine. Secondary doctrines, on the other hand, are those truths that should be believed in order to live rightly and receive the full benefit of God's benevolence, but about which there may be disagreement without threatening one's salvation. They are important, but not supremely important. They include such topics as church governance, the sacraments, predestination and free will, gender roles in the church and in the family, divorce and remarriage, sexuality, gifts of the Spirit, spiritual warfare, eschatology, and a host of social and political issues. In theory different views of secondary doctrines should not threaten Christian unity, but of course they often do.

Included among secondary doctrines are those that are relevant for the Revised Origins Paradigm: the age of the earth, creationism versus evolution, whether Adam and Eve were created morally perfect and were the only humans on earth at the time of the Fall, the nature of the Fall, and the specifics of the doctrine of original sin. While some Christians have elevated such doctrines to the level of a core doctrine or dogma, they have generally been held as secondary. They do not appear in the earliest statements of faith such as the Apostle's Creed (2nd Century) or Nicene Creed (4th Century). It is not until after the Reformation in the 16th and 17th Centuries that topics such as the Fall and original sin began to appear in various regional/denominational catechisms and confessions. However, such formal statements of faith generally included all doctrines, both core and secondary, and made no distinction between the two. They were, after all, intended to lay out all the proper beliefs of a given body of Christians, not to delineate the minimum that one must accept.

In addition to creeds, catechisms, and confessions, ecumenical councils have been the context in which many doctrines have been investigated, refined, and authorized. The so-called Jerusalem Council described in Acts Chapter 15 is generally considered to be the first such council, and there have twenty-one major councils since then (though only the Roman Catholic church recognizes the authority of all twenty-one). There have also been hundreds of minor councils and synods over the last two millennia which were called to address issues of limited scope or which were regional in nature, and were often advisory rather than conclusory. The Council of Carthage in 418 CE, at which Augustine's formulation of the doctrine of original sin was first accepted, was one of these synods.

It is important to note that the Council of Carthage in 418 CE was thoroughly Latin in character and language, and of the more than 200 attending bishops it is not clear if any were Greek. Located in northern Africa to the west of what is now Sicily, Carthage was situated within the Western, Latin half of the Christian world. While the Latin attendees of this minor council undoubtedly viewed their final declarations as being universally applicable to all Christians everywhere, Greek Christians did not.

For some time Greek and Latin Christianity had been developing nearly in isolation as a result of differences in language and politics, not to mention geography. While both sides considered themselves to be part of one and the same faith, their doctrines differed in a variety of areas. The two sides also took distinctly different approaches to resolving doctrinal disputes. While Latin Christians saw the bishop of Rome as the final authority, Greek Christians viewed the bishop of Rome as just one of an ecclesial collection of bishops from all regions who together were the ultimate authority in all church matters.

Greek Christians viewed mankind's inheritance from Adam as being mortality and corruption only, not his sin or guilt. In their view, Adam was the first sinner and the prototype of all sinners, and his disobedience resulted in him having a weakened (though not bankrupt) moral nature. But they saw the transmission of sin as occurring primarily through social systems rather than through biological heredity, and they placed a strong emphasis on humanity's free will and personal responsibility. In short, Greek Christians viewed every person as being responsible before God only for the sin he commits, not for the sin that Adam committed, though every person inherits an inclination to sin.

Arguably, Greek Christians should have had an advantage over Latin Christians when it came to understanding the meaning and intent of the New Testament since the original manuscripts were written in

Greek. The Latin theologians who championed the doctrine of original sin were working from Latin translations which had the potential for losing some of the nuances in the original Greek texts.

The differences between Latin and Greek views of original sin, while at times impassioned, did not result in a schism between the Latin and Greek churches in the 5th Century, nor in the centuries that followed. The two sides continued to exist in an uneasy spiritual kinship until the Great Schism which occurred in the 11th Century, and even then the schism was driven by issues other than original sin. In short, for hundreds of years Christians maintained communion with each other over core doctrines while maintaining different views over the secondary doctrines surrounding the Fall and original sin. This history demonstrates that it is possible to disagree on some doctrinal issues and still be part of the same family of God.

It remains for trained scholars and theologians to study the details of the Revised Origins Paradigm to determine what elements may need adjustment. That work has in fact already begun as the Revised Origins Paradigm was built upon the work of others. Until wider consensus is reached, it falls to each faithful Christian to determine what he considers most consistent with the facts as he knows and understands them.

Chapter 15
Christ Our Consolation

Every single person without exception has a need for Christ to make him whole spiritually, emotionally, mentally, and physically, and to reconcile him to God from Whom he has become estranged. The underlying reasons for this need may be somewhat different under the Revised Origins Paradigm than they are under the Traditional Origins Paradigm, but the fact of our brokenness and sinfulness is not. The doctrine of the cross remains the center of Christianity in every respect, without any alteration whatsoever.

That said, there is at least one aspect of Christ's work on the cross that, while no different under the Revised Origins Paradigm, may take on a heightened significance. This is His suffering as distinct and separate from His death. That He had to suffer is clear enough from the prophecies of the suffering servant in Psalm Chapter 22 and in Isaiah Chapters 52 and 53, not to mention Jesus's own statements to His disciples. What is less clear is what that suffering accomplished that His death could not.

The standard description of the purpose of the cross is that it resolves the spiritual crisis of man's separation from God as a result of sin and the consequent punishment that he deserves. At the center is the concept of penal substitution, Jesus taking on the punishment for man's sins. This act was a sacrifice and ransom which brought about reconciliation between God and man by satisfying the wrath of God incited by sin, making man righteous before God, and setting man on the path to healing in mind, body, and relationships.

> *God made him who had no sin to be sin for us, so that in him we might become the righteousness of God. (2 Corinthians 5:21)*

On the cross, Jesus became the consummate sin offering. The requirements for sin offerings under the law of Moses are provided in Leviticus Chapters 4 and 16, and dictate that a bull, goat, or lamb without defect be sacrificed for the sins of an individual or the community as a whole. The law explicitly states that such a sacrifice will be effective in atoning for sin, presumably through the faith of the offeror in the act prescribed by God.

> *In this way the priest will make atonement for them for the sin they have committed, and they will be forgiven. (Leviticus 4:35)*

Even so, animal sacrifices for sin fell short of what is necessary to fully repair what is broken in man and in the world; forgiveness removes guilt, but does nothing to bring about healing and restoration. Moreover, animal sacrifices addressed only unintentional sins. The only context in which more egregious, intentional sins could be addressed was through the scapegoat on Yom Kippur, the Day of Atonement.

> *He is to lay both hands on the head of the live goat and confess over it all the wickedness and rebellion of the Israelites - all their sins - and put them on the goat's head. He shall send the goat away into the wilderness in the care of someone appointed for the task. The goat will carry on itself all their sins to a remote place; and the man shall release it in the wilderness. (Leviticus 16:21-22)*

However, in this case the sins were removed only symbolically because there was no attendant sacrifice, and

> *without the shedding of blood there is no forgiveness. (Hebrews 9:22b)*

The blood of bulls and goats could provide temporary forgiveness for unintentional sins, but could not fully remove them. The scapegoat was intended to fully remove all sins, but only symbolically. A better sacrifice was needed to accomplish the full breadth of healing that God intended.

As the perfect sacrifice, Jesus was the "Lamb of God who takes away the sin of the world" (John 1:29). This act of "taking away" sins, foreshadowed in the symbolic scapegoat, resulted in the complete removal of sin so that it was no longer a factor in the situation. The blood of Jesus not only paid the penalty for man's sin, effecting forgiveness, but also rendered sin inoperative, totally annulling it. Moreover, it took an act of God to cleanse man's conscience and bring about transformation in his thoughts and behavior. The blood of animals could not accomplish that.

Jesus's sacrifice differs from animal sacrifices in multiple ways. While animals are free from sin and might on that basis appear to be suitable proxies for bearing man's sin, they have no understanding of the meaning of their sacrifice and cannot willingly offer themselves as a sacrifice for sin. The intentional sins of man require an intentional sacrifice as a remedy. Further, animals cannot act in obedience to God nor out of love for man. The removal of sins is not only a matter of satisfying God's justice, but also of God acting in love to free His creation from the deleterious effects of sin.

But perhaps the most significant difference between animal sacrifices and Jesus's sacrifice is in regards to suffering. Nowhere in the law of Moses is there any indication that an animal used as a sin offering was required to suffer; there are no prescriptions for torturing the animal prior to killing it to make the sacrifice effective or legitimate. In fact, both the Bible and Jewish tradition regard animals as deserving of a measure of respect and kindness.

The righteous care for the needs of their animals (Proverbs 12:10a)

God Himself is also portrayed as caring for animals.

You, LORD, preserve both people and animals. (Psalm 36:6b)

The suffering that Jesus endured is generally regarded as being the necessary result of Him assuming the punishment that was due to man for his sins.

> *But he was pierced for our transgressions, he was crushed for our iniquities; the punishment that brought us peace was on him, and by his wounds we are healed ... for the transgression of my people he was punished. (Isaiah 53:5, 8b)*

However, there are two aspects of Jesus's suffering that extend beyond mere substitution, man's punishment being transferred to Him. The first is that unlike the ultimate punishment that God inflicts on unredeemed man which is direct, God consigning him to a tortuous eternity in hell, the punishment that fell on Jesus was indirect, God abandoning Him into the hands of hateful and violent men. It is as if Jesus did not suffer punishment for sin so much as He suffered the natural consequence of sin. His sacrifice was to intentionally step into the sin of the world in its ugliest manifestation, letting sin have free reign to traumatize Him in body and soul. In so doing He absorbed, as it were, all of the ills that sin had wrought in the world. In the 4th Century Athanasius of Alexandria held a similar view, regarding Jesus as achieving His work of redemption as the representative of man rather than as his substitute.

Secondly, and most relevant to the Revised Origins Paradigm, Jesus's suffering went beyond punishment for the sins of mankind to include identification with the sorrows inherent in earthly life. In Isaiah Chapter 53, the very same chapter that portrays the suffering servant as taking on man's punishment, He is also portrayed as bearing all of man's pain regardless of its source.

> *Surely he took up our pain and bore our suffering (Isaiah 53:4a)*

Rather than pain and suffering, other translations use sicknesses, griefs, sorrows, weaknesses, and even disfigurements. These are not descriptions of sin (at least not directly) but rather the effects of living in a disordered world.

The Bible hints at God's pity for and unification with the suffering of His people in multiple places, such as in Matthew's story of the sheep and goats where He concludes with this:

The King will reply, "Truly I tell you, whatever you did for one of the least of these brothers and sisters of mine, you did for me." (Matthew 25:40)

Similarly, in the moment of Paul's first encounter with Jesus on the road to Damascus, Jesus says that what happens to God's people happens to Him.

"Saul, Saul, why do you persecute me?" (Acts 9:4b)

This truth is also evident in the Old Testament.

In all their distress he too was distressed (Isaiah 63:9a)

whoever touches you touches the apple of his eye (Zechariah 2:8b)

Along with Paul's description of the redemptive work of Christ in Romans Chapter 5, such passages led some early theologians to develop the concept of recapitulation wherein the incarnate Jesus was not merely the representative of mankind but in some mystical sense was one with them. According to Kelly, this concept arose with Irenaeus of Lyons in the 2nd Century who saw Christ as gathering together, including or comprising the whole of reality in Himself, including the human race. Similarly, Cyril of Alexandria said:

By virtue of the flesh united to Him, He has us all in Himself. (per Kelly)

The second Person of the Trinity entered earthly life not only to point the way to the Father and to die for man's sins, but also to demonstrate His oneness with man in all his earthly experience, in particular its adversities. He endured misunderstanding, hatred, betrayal, abandonment, mocking, condemnation, and of course physical pain of the most extreme kind. In the 2nd Century, Justin Martyr described it this way:

The Word of God became man for our sake, so that participating in our miseries He might heal them. (per Kelly)

Under the Revised Origins Paradigm, earthly existence necessarily includes suffering as a result of the disorder that has been an inherent aspect of the material universe since it was created. God began the process of establishing order and harmony in creation before handing the reigns to Adam and Eve to continue that work. They failed in their assigned vocation, and as a result that disorder increased rather than decreased, manifesting as both sin and suffering of all kinds. While Jesus demonstrated His rule over the natural order by calming the sea and by healing sicknesses during His time on earth, He did not eradicate all disorder. Perfect order, unity, and harmony in creation will not be established until God brings the new heavens and new earth at the end of the age.

> *See, I will create new heavens and a new earth. The former things will not be remembered, nor will they come to mind. (Isaiah 65:17)*

> *Then I saw "a new heaven and a new earth," for the first heaven and the first earth had passed away, and there was no longer any sea...He will wipe every tear from their eyes. There will be no more death or mourning or crying or pain, for the old order of things has passed away. (Revelation 21:1,4)*

Until then, God sympathizes with, and through the incarnation empathizes with all of man's sorrows. He neither explains nor removes suffering, but instead steps into it Himself and thereby proves His passionate love for man. The mystery of suffering, the bane of theologians and heathen philosophers alike for millennia, may remain unresolved for now, but we are given the grace to endure for a time through our abiding in Jesus Christ our Savior who shares in our suffering.

References

Bethune-Baker, J.F. *An Introduction to the Early History of Christian Doctrine: To the Time of the Council of Chalcedon*. London, England: Methune & Co. Ltd, 1920.

Biddle, Mark E. *Missing the Mark: Sin and its Consequences in Biblical Theology*. Nashville, Tennessee: Abingdon Press, 2005.

Boice, James Montgomery. *Foundations of the Christian Faith: A Comprehensive & Readable Theology*. Downers Grove, Illinois: InterVarsity Press, 1986.

Bouteneff, Peter C. *Beginnings: Ancient Christian Readings of the Biblical Creation Narratives*. Grand Rapids, Michigan: Baker Academic, 2008.

Calvin, John (Edited by Tony Lane and Hilary Osborne). *The Institutes of Christian Religion.* Grand Rapids, Michigan: Baker Book House, 1987.

Cole-Turner, Ron. *The End of Adam and Eve: Theology & the Science of Human Origins*. Smashwords, 2016.

Collins, C. John. *Did Adam and Eve Really Exist? Who They Were and Why You Should Care*. Wheaton, Illinois: Crossway, 2011.

Constantelos, Demetrios J. *Understanding the Greek Orthodox Church: Its Faith, History and Life*. Brookline, Massachusetts: Hellenic College Press, 2005.

David, Rabbi Boruch. *Adam and Eve The Rest of the Story: Secrets from the Ancient Wisdom of the Jews*. Xlibris, 2016.

Enns, Peter. *The Evolution of Adam: What the Bible Does and Doesn't Say About Human Origins*. Grand Rapids, Michigan: Brazos Press, 2012.

Ferguson, Everett. *Baptism in the Early Church: History, Theology, and Liturgy in the First Five Centuries*. Grand Rapids, Michigan: William B. Eerdmans Publishing Company, 2009.

Gaebelein, Frank E. *The Expositor's Bible Commentary*. Grand Rapids, Michigan: Zondervan Publishing House, 1986.

Gonzalez, Justo L. *The Story of Christianity: Volume I, The Early Church to the Dawn of the Reformation*. New York, New York: HarperCollins Publishers, Inc., 2010.

Holmes, Jean. *Do Dogs Go to Heaven? Eternal Answers for Animal Lovers*. Joipax, Inc. via Smashwords, 2015 (3rd edition).

Irenaeus. *The Proof or Demonstration of the Apostolic Preaching*. CrossReach Publications, 2017.

Jacobs, Alan. *Original Sin: A Cultural History*. New York, New York: HarperCollins Publishers, Inc., 2008.

Joyner, Rick. *There Were Two Trees in the Garden*. New Kensington, Pennsylvania: Whitaker House, 1992.

Kelly, J.N.D. *Early Christian Doctrines*. HarperSanFrancisco, 1978 (fifth edition).

King, Robert Alan. *A Christian Rebuttal to Steven Woodward's Biblical Proof Animals Do Go To Heaven: The Real Truth About Dogs, Cats, Pets, and the Afterlife*. King & Associates via Smashwords, 2013.

Kriwaczek, Paul. *Babylon: Mesopotamia and the Birth of Civilization*. New York, New York: St. Martin's Press, 2012.

Leick, Gwendolyn. *Mesopotamia: The Invention of the City*. New York, New York: Penguin Books, 2001.

Lewis, C.S. *The Problem of Pain*. New York, New York: Macmillan Publishing Company, 1978 (21st printing).

Louth, Andrew. *Ancient Christian Commentary on Scripture (I: Genesis 1-11)*. Downers Grove, Illinois: InterVarsity Press, 2001.

Mook, James R. "*The Church Fathers on Genesis, the Flood, and the Age of the Earth*". See Chapter 1 of Mortenson and Ury.

Morgan, Christopher W., Peterson, Robert A. *Suffering and the Goodness of God.* Wheaton, Illinois: Crossway Books, 2008.

Morris, Henry M. *Biblical Creationism: What Each Book of the Bible Teaches About Creation and the Flood.* Green Forest, Arizona: Master Books, 2008 (4th printing).

Mortenson, Terry; Ury, Thane H. *Coming To Grips With Genesis: Biblical Authority and the Age of the Earth.* Green Forest, Arizona: Master Books, 2012 (6th printing).

Osborn, Ronald E. *Death Before the Fall: Biblical Literalism and the Problem of Animal Suffering.* Downers Grove, Illinois: InterVarsity Press, 2014.

Pelikan, Jaroslav. *The Christian Tradition: A History of the Development of Doctrine; 1 The Emergence of the Catholic Tradition (100 - 600).* Chicago, Illinois: The University of Chicago Press, 1975 (paperback edition).

Philippe, Jacques. *Interior Freedom.* New York: Scepter Publishers, Inc., 2007 (English translation).

Pusey, Edward B. *The Confessions of Saint Augustine.* New York, New York: Macmillan Publishing Company, 1961.

Quasten, Johannes. *Patrology (Volume IV: The Golden Age of Latin Patristic Literature from the Council of Nicea to the Council of Chalcedon).* Allen, Texas: Christian Classics, 1986.

Quiggle, James D. *Adam and Eve: A Biography and Theology.* James Quiggle via Smashwords, 2012.

Rose, Fr. Seraphim. *The Place of Blessed Augustine in the Orthodox Church.* Platina, California: Saint Herman Press, 2007 (3rd edition).

Russell, Norman. *Cyril of Alexandria.* New York, New York: Routledge, 2000.

Santillana, Giorgio de. *The Crime of Galileo*. University of Chicago Press, 1955.

Sproul, R.C. *Chosen By God*. Carol Stream, Illinois: Tyndale House Publishers, Inc., 1986.

Stambaugh, James. *"Whence Cometh Death? A Biblical Theology of Physical Death and Natural Evil"*. See Chapter 13 of Mortenson and Ury.

Strickland, Ashley. *"Modern fossil discovery rewrites human history"*. Cable News Network (CNN), January 25, 2018.

Toews, John E. *The Story of Original Sin*. Eugene, Oregon: Pickwick Publications, 2013.

Vanneste, Alfred. *The Dogma of Original Sin*. Published in Brussels by Vander; Published in Louvain by Nauwelaerts. 1975. As cited by Wiley.

Vincent of Lerins (edited by Reginald S. Moxon). *The Commonitorium of Vincentius of Lerins*. London, England: Cambridge University Press, 1915.

Wallace, A.J., Rusk, R.D. *Moral Transformation: The Original Christian Paradigm of Salvation*. Seattle, Washington: CreateSpace, 2011.

Walton, John H. *The Lost World of Adam and Eve: Genesis 2-3 and the Human Origins Debate*. Downers Grove, Illinois: InterVarsity Press, 2015.

Whorton, Mark S. *Peril in Paradise: Theology, Science, and the Age of the Earth as cited by Osborn.*

Wiley, Tatha. *Original Sin: Origins, Developments, Contemporary Meanings*. Mahwah, New Jersey: Paulist Press, 2002.

David Korotney has degress in Chemical and Mechanical Engineering and worked for the Environmental Protection Agency for 32 years before retiring in 2023. He and his wife Abby live in Ypsilanti, Michigan and have two grown sons.

www.ingramcontent.com/pod-product-compliance
Lightning Source LLC
LaVergne TN
LVHW052255200926
848481LV00002B/377
9798230036180